Chameleons
OF SOUTHERN AFRICA

Krystal Tolley &
Marius Burger

Published by Struik Nature
(an imprint of Penguin Random House SA (Pty) Ltd)
Reg. No. 1953/000441/07
Estuaries No.4, Oxbow Crescent, Century Avenue,
Century City, 7441
PO Box 1144, Cape Town, 8000, South Africa

Visit www.penguinrandomhouse.co.za and join the Struik
Nature Club for updates, news, events and special offers.

First published in 2007
10 9 8 7 6 5

Copyright © in text, 2007: Krystal Tolley
& Marius Burger
Copyright © in photographs, 2007: Krystal Tolley
& Marius Burger, or as indicated alongside images
Copyright © in illustrations, 2007: Penguin Random
House SA (Pty) Ltd
Copyright © in maps, 2007: Penguin Random House
SA (Pty) Ltd
Copyright © in published edition, 2007: Penguin
Random House SA (Pty) Ltd

Publishing manager: Pippa Parker
Managing editor: Rod Baker
Editor: Helen de Villiers
Designer: Janice Evans
Cartographers: Rene Navarro of UCT's
Avian Demography Unit (distribution maps);
James Whitelaw (maps pp. 8, 21; chart p. 27)
Illustrator: Simon van Noort (image p. 46-7)
Proofreader: Glynne Newlands
Indexer: Cora Ovens

Reproduction by Hirt & Carter Cape (Pty) Ltd
Printed and bound by RR Donnelley Asia Printing Solutions
Ltd., Hong Kong

All rights reserved. No part of this publication may be
reproduced, stored in a retrieval system, or transmitted,
in any form or by any means, electronic, mechanical,
photocopying, recording or otherwise, without the prior
written permission of the copyright owner(s).

ISBN 978 1 77007 375 3
EPUB 978 1 92057 286 0
EPDF 978 1 92057 287 7

PICTURE CREDITS

All photographs are by Krystal Tolley and Marius
Burger, except where indicated differently next to
photographs. Contributing photographers are Chris
Anderson, Randy Babb, Bill Branch, Hugh Chittenden,
Michael Cunningham, Atherton de Villiers, Jonathan
Losos, Bill Love, Johan Marais, Adnan Moussalli, Ben
Phillips, Devi Stuart-Fox, Colin Tilbury, Simon van
Noort, Miguel Vences, John Visser and Bart Wursten.

Front cover: Cape Dwarf Chameleon
Title page: Juvenile Eastern Cape Dwarf Chameleon
Below: Two different forms of the
Wolkberg Dwarf Chameleon
Opposite, top to bottom: Flap-necked Chameleon;
Namaqua Chameleon; Mulanje Pygmy Chameleon;
Cape Dwarf Chameleon;
Undescribed pygmy chameleon
Back cover: Common Flap-necked Chameleon

CONTENTS

Preface 4

Acknowledgements 5

SECTION ONE 6

Distribution and classification 8

Evolution and the fossil record 26

Ecology and anatomy 29

Behaviour and reproduction 34

The amateur naturalist 40

SECTION TWO 44

Species identification 46

   Dwarf chameleons – *Bradypodion* 48

   Undescribed dwarf chameleons 80

   Typical chameleons – *Chamaeleo* 86

   Pygmy chameleons – *Rhampholeon* 90

Glossary 94

Resources 96

Index 98

Words in **bold** are defined in the glossary on pages 94–5.

PREFACE

This guide was written in response to popular demand. In the course of our work on chameleons, we have often received phone calls and emails from people asking basic questions about these creatures. Most start off with 'I have a chameleon living in my garden, and I was wondering …' Not that we became tired of answering emails, or chatting on the phone with people from Polokwane to Cape Town, but it occurred to us that there was little literature readily available for those with a specific interest in chameleons. We came to realise that it was time to take the leap, and we began to plan out a short field guide for the layperson. It didn't take much planning, as we simply had to recall all the questions we've been asked about chameleons and set about recording the answers. That's how and why this field guide was created … from the very questions that you have been asking.

The biggest challenge was compiling the species accounts. One of the reasons why it is so hard to tell chameleons apart is that different species can often look very similar (even to us, the so-called experts), and, at the same time, the appearance of different members of one species can be incredibly varied. Finding particular characteristics that distinguish the species has long been a frustrating problem for researchers and naturalists alike. There are few consistencies of identification characteristics, and we admit that we sometimes initially identify a chameleon by means of 'gut feeling'. Regardless, this guide was prepared in a way that should help the amateur naturalist to identify chameleons in southern Africa, and to answer many of those questions posed in your phone calls and emails.

KRYSTAL TOLLEY & MARIUS BURGER
Cape Town, 2007

American-born **Krystal Tolley** is head of the Molecular Ecology and Evolution Programme at the South African National Biodiversity Institute at Kirstenbosch. She received her PhD in Norway, originally working on marine mammals. Since moving to South Africa, she has published numerous scientific and popular articles on chameleons. She has spent many days in the field searching for chameleons across southern Africa, and has a keen interest in understanding their biodiversity.

Marius Burger became interested in reptiles and amphibians at a young age. After 11 years with Eastern Cape Nature Conservation, he turned freelance to conduct herpetological surveys in southern and central Africa. His favourite hangout is Madagascar, where he leads natural history tours. He is currently employed as Project Herpetologist for the Southern African Reptile Conservation Assessment, a project of the South African National Biodiversity Institute.

ACKNOWLEDGEMENTS

Creating a book like this can only be a team effort, with participants ranging from those who have contributed photos, unpublished data and anecdotal information, to those who simply talked with us about chameleons. Most notably in this regard are Bill Branch, Devi Stuart-Fox and Colin Tilbury – may they live long and happy lives. For those who have assisted with fieldwork, often spending countless cold and hungry hours searching for chameleons with us, your hard work has made all the difference. Special thanks go to Michael Cunningham, Kate Henderson, Andrew Turner and Simon van Noort. Some folk assisted with logistics, proofreading or provided valuable information used in this guide. Our thanks go to Adrian Armstrong, Daphne and Bruce Beyer, Derek Clark, Michael Cunningham, Sarah Davies, Atherton de Villiers, Dahné du Toit, Vincent Egan, Kate Henderson, Elton le Roux, Georgio Lombardi, John Manning, Johan Marais, Adnan Moussalli, le Fras Mouton, Carel Oosthuisen, Alvin Page, Les Powrie, Mike Rutherford, Gordon Setaro, Rodger Smith, Samantha Stoffberg, Belinda Swart, Ernst Swartz, Jerry Theron, Andrew Turner, Jenny Underhill, Pierre van den Berg, Simon van Noort, John Visser and Kelley Whitaker. Scores of field rangers and managers of provincial reserves have gone beyond the call of duty to facilitate our research efforts. Uncountable private landowners have allowed us access to their properties, and much of the information in this field guide would have remained unknown were it not for their assistance.

We thank Graham Alexander (University of Witwatersrand) for proofreading and commenting on the entire manuscript. Most of Krystal Tolley's earlier chameleon research was conducted through the University of Stellenbosch Evolutionary Genomics Group. Thanks to Conrad Matthee for providing the opportunity to conduct this research, and for obtaining a grant from the National Research Foundation to partially fund this research. Fieldwork was partly funded by the South African National Biodiversity Institute (SANBI). Rene Navarro at the Avian Demography Unit (ADU), University of Cape Town, designed the distribution maps. Some of the distribution data was provided by Bayworld (Port Elizabeth Museum), South African Museum, Transvaal Museum, National Museum, Natal Museum, Durban Museum, and the Southern African Reptile Conservation Assessment (SARCA). Finally, we'd like to thank the team at Struik – Pippa for believing in the idea, and Janice and Helen for making it happen.

A fly falls prey to a Beardless Dwarf Chameleon.

SECTION ONE

Chameleons are among the world's most remarkable lizards. They inspire wonder and fascination among nature-lovers, both amateur and professional. By contrast, in some human cultures, they evoke mystery and even fear. The aura surrounding chameleons, good or bad, is most likely a result of their strange and unusual characteristics: they change colour, have a projectile tongue for catching prey, swivel their eyes in all directions in a rather unnerving manner, have a peculiar gait, and usually sit with their tail curled into an impressive spiral. Yet there is much more to be discovered about this interesting group of lizards, and the exploration begins here with an overview of all types of chameleons (of which there are up to 160 species), their history, geographic distribution (restricted mainly to Madagascar and Africa), reproduction, behaviour and their relationships to one another.

Spiralled tail of a Cape Dwarf Chameleon.

DISTRIBUTION AND CLASSIFICATION

Chameleons are not found worldwide, but have a distribution that is restricted mainly to Madagascar, Africa and some of the neighbouring islands (e.g. Comoros, Mauritius, Seychelles and Zanzibar), as well as the extreme southern fringes of Europe, the Arabian Peninsula and India (see map below). There are approximately 150 to 160 **species**, which can be arranged into nine groupings (**genera**) based both on **morphological** and **genetic** characteristics. Three genera (*Brookesia, Calumma* and *Furcifer*) are found almost exclusively in Madagascar, with a few species of *Furcifer* also on neighbouring Indian Ocean islands. The **genus** *Bradypodion* is endemic to South Africa, and *Rhampholeon* and *Rieppeleon* are found in East, Central and southern Africa.

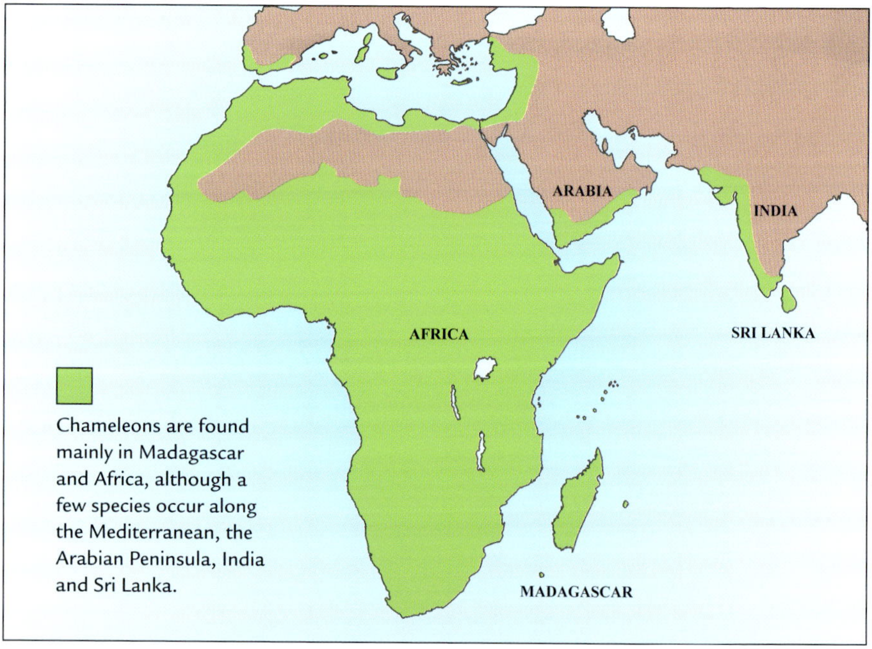

Chameleons are found mainly in Madagascar and Africa, although a few species occur along the Mediterranean, the Arabian Peninsula, India and Sri Lanka.

C Tilbury

In addition, two new genera, *Kinyongia* from East Africa and *Nadzikambia* from Malawi, were recently named. *Chamaeleo* is the most widespread genus: it is distributed across much of Africa and contains the only four species that are also found outside Africa and Madagascar. These four species are not widespread, occurring only along the Mediterranean coastal areas – Mediterranean Chameleon (*Chamaeleo chamaeleon*); the margin of the Arabian Peninsula and the coast of India – Arabian Chameleon (*Chamaeleo arabicus*) and Veiled Chameleon *(Chamaeleo calyptratus*); and Sri Lanka – Asian Chameleon (*Chamaeleo zeylanicus*).

Chameleons occur in many types of habitat from grasslands to deserts, forest canopies and forest floors. Some species can be found along coastlines and have even been observed wandering in the intertidal zone, while other species occur on the slopes of some of the highest mountains; for example, chameleons have been found as high as 4 500 m elevation on the slopes of Mt Kenya.

B Love

D Stuart-Fox

Left: Adult and juvenile Veiled Chameleons.
Above: Mt Kenya is a volcanic cone that rises 5 200 m above sea level. At least four species of chameleon occur on its high, forested slopes – *Kinyongia exubitor, Chamaeleo hoehnelii, C. jacksonii* and *C. schubotzi*.
Opposite and below, left to right: Six genera occur in Africa: *Bradypodion, Chamaeleo, Kinyongia, Nadzikambia, Rhampholeon* and *Rieppeleon*.

C Tilbury

C Tilbury

C Tilbury

CHAMELEONS OF MADAGASCAR

Madagascar is well known as a biodiversity hotspot. This island, which is almost half the size of South Africa, contains a rich and unique variety of insects, mammals, birds, frogs, snakes and lizards. Chameleons are particularly well represented, with more than half of the world's species found only in Madagascar. Some of the largest chameleons occur there, including Parson's Chameleon (*Calumma parsonii*) and Oustalet's Chameleon (*Furcifer oustaleti*), both of which measure nearly 70 cm in total length. Madagascar also claims the smallest chameleons, the Nosy Bé Leaf Chameleon (*Brookesia minima*) and the Warty Leaf Chameleon (*Brookesia tuberculata*), which barely reach 3 cm in total length.

Madagascar boasts the world's largest and smallest chameleons: the gigantic Parson's Chameleon (*Calumma parsonii*), and the diminutive Warty Leaf Chameleon (*Brookesia tuberculata*).

Three genera of chameleon occur on Madagascar: *Brookesia*, *Furcifer* and *Calumma*. *Brookesia*, or Malagasy leaf chameleons, are typically ground dwelling (**terrestrial**), living in the leaf-litter on the forest floor by day, and climbing onto low vegetation to sleep at night. All are very **cryptic**, being brownish or greenish, with only a limited ability to change colour. Their short tails and compressed sides render them leaf-like in appearance. The leaf-mimicry is very convincing, to the extent that some are patterned with leaf **venations**. Several species are covered in fleshy bristles, which serve to break the outline of the animal as it clambers through the leaf-litter. Some are found in, and resemble, the mosses that grow on trees.

It was once thought that the Malagasy leaf chameleons (*Brookesia*) were related to the African pygmy chameleons (*Rhampholeon*) because they are seemingly so similar. Both are somewhat leaf shaped, with stumpy tails. Genetic studies have now confirmed that they are only distant relatives, and we now know that their superficial resemblance

Morphological variation in Malagasy leaf chameleons: *Brookesia ramanantsoai*, *B. stumpffi*, *B. perarmata*.

is due to **convergent evolution**. This refers to instances where unrelated species have independently adapted to the similar environments in which they live, and so evolved to look similar. An example of convergent evolution is the flippers of whales and the fins of fish. Both live in the same environment and have adaptations for an aquatic lifestyle (flippers and fins), but this similarity does not mean that they are closely related. For *Brookesia* and *Rhampholeon*, the adaptation of a body form shaped like a leaf has become prevalent because it allows them to blend in easily with the leaf-litter of the forest floors they both inhabit, and this camouflage is the key to their survival.

M Vences

Some Malagasy leaf chameleons, such as Vadon's Leaf Chameleon (*Brookesia vadoni*), easily blend into a background of mosses.

Both *Brookesia superciliaris* and *Rhampholeon spectrum* inhabit the leaf-litter of the forest floor. They look very similar as a result of convergent evolution.

Chameleons of the genus *Calumma* are found on Madagascar, with two species on the Comoros and the Seychelles. On Madagascar, they tend to inhabit the wetter forests and mountainous regions. Mature *Calumma* males are larger than females, but the various species range in size from the small, barely longer than 10 cm, Nose-horned Chameleon (*C. nasuta*), to one of the world's giants, the Parson's Chameleon (*C. parsonii*) which exceeds 60 cm in total length. *Calumma* are most famous for their elaborate, gaudy nasal appendages, which give them a Pinocchio-like appearance. The males of some species have nasal appendages that are paired or even forked. Several of the *Calumma* species also have **occipital lobes** or skin-flaps that are ear-like in appearance, and these lobes are similar to, but generally larger than, those of the Common Flap-necked Chameleon that occurs in southern Africa.

Morphological variation within the Malagasy genus, *Calumma*: *C. gastrotaenia* (above left) and *C. brevicornis* (above right). *C. nasuta* (below) is the smallest species in the genus.

Species of *Furcifer* come in all shapes, sizes and colours. Some exhibit extreme **sexual dimorphism**, and males and females of the same species have a very different appearance from one another. Males can be substantially larger than females, and some species have nasal appendages.

Two species of the genus *Furcifer*: *F. pardalis* (left) and *F. campani* (right).

Extreme sexual dimorphism occurs in *Furcifer minor* – the male is on the left.

Two species of *Furcifer* are found on the Comoros, but the majority occur on Madagascar where they are widely distributed. Most inhabit the arid western and southern regions; however, a few species are found in the central mountains as well as the wetter habitats of the east. Oustalet's Chameleon (*F. oustalei*) is widespread and found in all these habitats. Although Madagascar suffers from a great deal of habitat transformation, *Furcifer* seem to be able to survive in degraded habitats and in towns, where they are easily spotted.

True or false chameleons?

There has long been a tendency to lump chameleon genera into two larger groupings: 'true chameleons' and 'false chameleons'. True chameleons are called the Chamaeleoninae and consist of the genera *Bradypodion*, *Calumma*, *Chamaeleo* and *Furcifer*. False chameleons are called the Brookesiinae, and include the genera *Rhampholeon* and *Brookesia*. The grouping implies that the members within each group are closely related to one another. While these classifications persist, recent evidence shows that they are not valid. Genetic studies indicate that the genera within these groupings are not necessarily closely related to one another and should not be grouped together into 'true' and 'false' chameleons.

The mechanics of classification

When new **species** are discovered, they are given a scientific name. For example, the scientific name for the human species is *Homo sapiens*. Each name (binomen) consists of the 'genus', as in *Homo*, and the 'specific epithet', as in *sapiens*. Together, these terms make up the scientific name of the species, which is always written in italics. If the genus name has already been mentioned in a text, it is abbreviated in subsequent usage; in this case the binomen would be *H. sapiens* after the initial usage.

All organisms are grouped into a scientific classification system. The most inclusive groups are the **Kingdoms**, of which the animals (Animalia) are but one Kingdom of six (the others are Plantae, Fungi, Protista, Archaebacteria and Eubacteria). Each Kingdom is made up of several **Phyla** (singular **Phylum**), which are themselves made up of several **Classes**, then **Orders**. Within Orders are **Families**, and all chameleons are found in the Family Chamaeleonidae. Closely related species within a family are then grouped into the same genus (plural genera). Scientific classification is a foolproof way for everybody, regardless of their location or language, to keep tabs on the who's who of life.

In the species accounts of this field guide, the species name is followed by the name of the author who first described the species, and the year in which this occurred. When the author and date are in brackets, it means that the

An example of classification for the Drakensberg Dwarf Chameleon:

- **Kingdom:** Animalia (animals)
- **Phylum:** Chordata (animals with a spinal cord)
- **Class:** Reptilia (reptiles)
- **Order:** Squamata (lizards and snakes)
- **Family:** Chamaeleonidae (chameleons)
- **Genus:** *Bradypodion* (dwarf chameleons)
- **Species:** *dracomontanum* ('dragon' & 'mountain')
- *Bradypodion dracomontanum* = Drakensberg Dwarf Chameleon

name has changed from that originally given. For instance, the first few dwarf chameleons (*Bradypodion*) described used to be in either the genus *Lophosaura* or *Microsaura* and those names will be in brackets.

Common names

The common English name 'chameleon' comes from the Greek *khamaileon* which is a combination of *khamai* meaning 'on the ground' (also 'dwarf') and *leon* meaning 'lion'. Essentially then, chameleons are 'ground-lions' or 'dwarf-lions'. The name of the genus, *Chamaeleo*, has the same derivation. *Bradypodion* comes from the Latin for 'slow-footed'. The Xhosa call chameleons *ilovane*; they are *lefokolodi* or *lihobu* in Sotho; *luhavihavi* in Venda; *rimpfana* in Tsonga; *unwabu* in Zulu; and the Swahili word for chameleon is *kinyonga*. In Afrikaans, the common names for chameleons are *verkleurmannetjie* meaning 'colourful little man', or *trapsuitjies* which means 'treading carefully'.

Courtesy of the Linda Hall Library of Science, Engineering and Technology, Kansas, USA.

Above: *Caméléon du Cap de Bonne Esperance*, plate from Tachard's *Voyage de Siam* (1686). A rather fanciful representation of the Namaqua Chameleon.
Left: The real Namaqua Chameleon.

Above left: A male Jackson's Three-horned Chameleon (*Chamaeleo (Trioceros) jacksonii*) from Mount Kenya.
Above right: The newest *Rhampholeon* is the enigmatic Usambara Spiny Pygmy Chameleon *(Rhampholeon spinosum).*

CHAMELEONS OF AFRICA

The **taxonomy** of African chameleons is still being researched, and future name changes are inevitable. Currently, there are six genera of chameleons in Africa. The *Bradypodion* (dwarf chameleons) are 'proudly South African', and will be discussed in detail in this field guide. The *Chamaeleo* (typical chameleons) are the most widespread group and are found throughout most of Africa. In southern Africa, there are two species of *Chamaeleo* and they are discussed in more detail in the species accounts.

The majority of species in the genus *Chamaeleo* are usually regarded as being in the sub-genus *Trioceros* (three-horned chameleons). There are nearly 40 species of

The Cameroon Sailfin Chameleon (*Chamaeleo montium*) is characterised by having an enlarged and flattened dorsal crest and tail.

An undescribed species of pygmy chameleon from East Africa.

Trioceros and they are fairly widespread across East and West Africa. They are medium-sized chameleons (20–30 cm total length) and some species have remarkable horns projecting from the snout. Perhaps even more impressive is the sail-shaped fin on the back and tail of the males of a few species, such as the Cameroon Sailfin Chameleon (*Chamaeleo* (*Trioceros*) *montium*) and the Four-horned Chameleon (*Chamaeleo* (*Trioceros*) *quadricornis*) of West Africa. This 'fin' is essentially skin that is stretched over elongated vertebral spines.

The **taxonomy** of the three-horned chameleons is currently being studied, but it appears that they are actually a distinct group that is only distantly related to *Chamaeleo*. Nevertheless, they are not yet considered to be their own genus, and for the time being, they remain under the *Chamaeleo*.

The *Rhampholeon* (African pygmy chameleons) are found in both East and West Africa, and reach their southern distribution limit in Zimbabwe. There are several species in this genus, most of which are restricted to **montane** forests in East Africa (e.g. *Rhampholeon moyeri, R. boulengeri, R. platyceps, R. chapmanorum, R. nchisiensis* and *R. uluguruensis*). Most species have very restricted distributions, the exceptions being *R. spectrum* which is found across parts of West Africa, and *R. temporalis* which has a wide distribution in East Africa. Marshall's Pygmy Chameleon (*R. marshalli*) and the Gorongoza Pygmy Chameleon (*R. gorongosae*) are found only in isolated areas of Zimbabwe and Mozambique respectively. A new species, the Usambara Spiny Pygmy Chameleon (*R. spinosum*), was recently transferred to this genus.

Two of the three *Rieppeleon*: the Short-tailed Pygmy Chameleon (*R. brevicaudatus*) and the Kenya Pygmy Chameleon (*R. kerstenii*).

Rieppeleon is a new genus, first named in 2004, of just three species: *Rieppeleon brachyurus*, *R. kerstenii* and *R. brevicaudatus*. These species have large and sometimes overlapping distributions in East and West Africa. They were formerly thought to be within *Rhampholeon*, but genetic and morphological studies have shown that they are a separate group that is no more related to *Rhampholeon* than they are to any other chameleons.

Two new genera of chameleons were named in 2006 – *Kinyongia* and *Nadzikambia*. *Kinyongia* are generally found in East Africa, and were formerly included in *Bradypodion*. However, DNA testing has revealed that they are not closely related to *Bradypodion*, and thus a new genus was created for them. There are nine species in this group – *K. adolfifriderici*, *K. carpenteri*, *K. excubitor*, *K. fischeri*, *K. tavetanum*, *K. uthmoelleri*, *K. xenorhinum*, *K. oxyrhinum* and *K. tenue*. Although most occur in East Africa, *K. adolfifriderici* is found in the Democratic Republic of the Congo and *K. excubitor* on Mt Kenya. All *Kinyongia* species occur in isolated tropical or subtropical forests.

The Usambara Two-horned Chameleon (*K. fischeri*) is one of nine members in the genus *Kinyongia*.

The sole member of the genus *Nadzikambia* is the Mt Mulanje Chameleon (*N. mlanjense*).

Some species (*K. fischeri* and *K. tavetanum*) have forked horns, whereas others (*K. carpenteri, K. xenorhinum, K. tenue* and *K. oxyrhinum*) have a single horn.

There is only one species in the new genus *Nadzikambia*, the Mt Mulanje Chameleon (*Nadzikambia mlanjense*). True to its name, this chameleon is endemic to Mt Mulanje in Malawi. It differs from other chameleons in both its morphology and its genetic make-up, and has no known close relatives.

What's new?

Taxonomists are scientists who name species and assign them to groups of closely related species. Early taxonomists based their work on morphological characteristics, but in many cases, species that are unrelated can look and behave in similar ways. This has caused many cases of distantly related species to be incorrectly grouped together. Taxonomists have recently added DNA to their box of tools to help them name and classify species. DNA studies suggest that many species of chameleon have been classified incorrectly, and taxonomists are endeavouring to reclassify them. Recently, three species of Rhampholeon were found to be different from all other Rhampholeon species, a finding that was confirmed by DNA analysis. Consequently, a new genus *Rieppeleon* was named for these species. Even more recently, two new genera were named (*Kinyongia* and *Nadzikambia*). The expectation is that several other new species and genera may be named in the future.

BIOMES OF SOUTHERN AFRICA

An appreciation of the physical environment is essential for understanding how and where the various chameleon species are distributed in southern Africa. The region covered in this book includes South Africa, Namibia, Botswana, Lesotho, Swaziland, Zimbabwe and Mozambique (as far north as the Zambezi River). Vegetation in this greater region is commonly divided into nine major biomes – fynbos, succulent Karoo, Nama Karoo, forest, Albany thicket, the Indian Ocean coastal belt, savanna, grassland and desert. The biomes are defined by the dominant plant growth forms. The number of chameleon species differs between the various biomes, with some being much more species rich than others.

The fynbos biome is a major component of the Cape Floristic Region, one of the world's most important biodiversity hotspots.

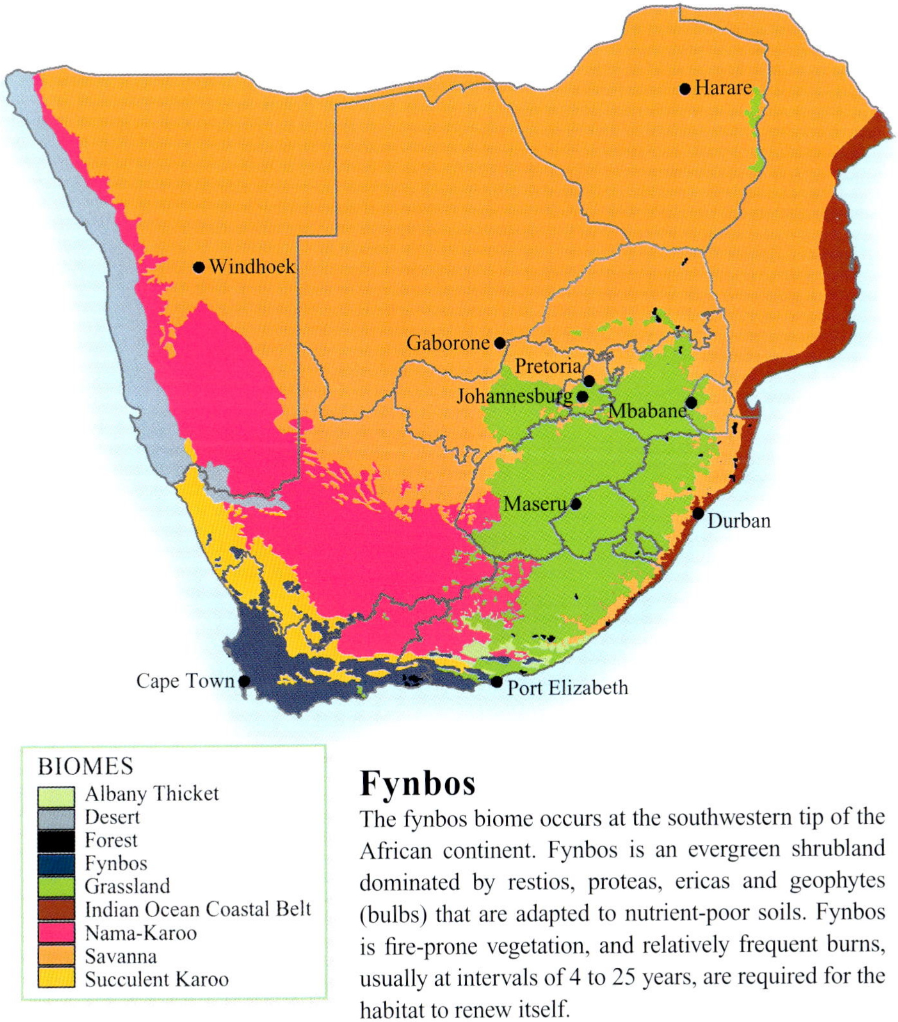

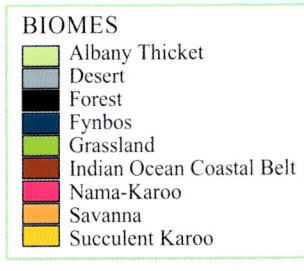

BIOMES
☐ Albany Thicket
☐ Desert
☐ Forest
☐ Fynbos
☐ Grassland
☐ Indian Ocean Coastal Belt
☐ Nama-Karoo
☐ Savanna
☐ Succulent Karoo

Fynbos

The fynbos biome occurs at the southwestern tip of the African continent. Fynbos is an evergreen shrubland dominated by restios, proteas, ericas and geophytes (bulbs) that are adapted to nutrient-poor soils. Fynbos is fire-prone vegetation, and relatively frequent burns, usually at intervals of 4 to 25 years, are required for the habitat to renew itself.

Compared to the interior of the region, the fynbos biome is an area of relatively high rainfall, receiving moderate winter rainfall in the west and year-round rainfall in the east. There are three species of chameleon in this biome, but recent research indicates that the **species richness** may be more than double this figure.

The fynbos biome is the predominant vegetation zone in the Cape Floristic Region (CFR), an area that extends from the Bokkeveld Mountains southwards to the coastal plain, and then eastwards to Port Elizabeth. Smaller areas of indigenous forest, thicket and succulent Karoo biomes make up the rest of the vegetation in the CFR. The CFR is a major hotspot of biological diversity, boasting more than 9 000 different plant species, 7 000 of which are **endemic**. There are at least five species of chameleon in the CFR, but this number is approximately double if undescribed species are included.

Succulent Karoo

The succulent Karoo biome runs along the northwest coast of South Africa before turning inland. It is more arid than fynbos, receiving less than 200 mm of rain each winter. The cold Benguela Current that runs northwards along the coast contributes to the aridity by inhibiting precipitation over the region. The area contains a wide diversity of succulent plants and annual flowers, known to many for their mass floral displays in springtime, when the veld becomes covered in a blanket of colours. The biome is a hotspot for biodiversity, with more than 2 400 endemic plants. Much of the succulent Karoo falls within the geographic region known as Namaqualand. Only two species of chameleon occur in the succulent Karoo.

Walter Knirr/IOA

Nama Karoo

The Nama Karoo receives low rainfall, mainly in late summer, and as a result is characterised by a thin cover of low shrubs and succulent plants. This biome stretches across the geographic region known as the Upper Karoo, a high plateau that occupies most of the interior of South Africa. Chameleons are scarce in the Upper Karoo, with only two species known from this region. Reports of other chameleons are focused around towns and settlements and may actually represent translocated individuals rather than naturally occurring species.

Hein Von Horsten/ IOA

Forest

The forest biome occurs in discontinuous patches along the Cape Fold Mountains, the eastern escarpment and along the east coast. There are two main types of forest – Afromontane forest (temperate) and Indian Ocean coastal forest (subtropical). Afromontane forests are scattered along the eastern escarpment southwards to the Cape Fold Mountains. The largest remaining fragment, the Knysna Forest, is found along the southern Cape coast, with a second substantial remnant patch further west at Grootvadersbosch. Forest chameleons are found in both these relatively large forest patches. Many of the ravines along the south-facing, wet slopes of the Outeniqua and Tsitsikamma mountains also have small patches of forest, but these fragments are too small to support forest chameleons. Afromontane forest patches are also scattered northwards along the eastern slopes of the Drakensberg into Limpopo, and forest-living chameleons are known from many of these patches. Indian Ocean coastal forests are generally found in the coastal zone of the Eastern Cape, where they are fragmented in distribution; they become more continuous where they extend into KwaZulu-Natal. These forests form a large part of the Indian Ocean coastal belt biome.

Forests once dominated the southern African landscape during wetter and warmer periods of the Earth's history. Today, southern African forests are but relics. They have slowly shrunk in size, having been replaced by more arid biomes as the climate has become progressively cooler and drier over the last tens of millions of years. Thus, forest-adapted chameleons once had a larger area in which to roam, but populations have since become isolated in small, separate patches. Currently, these isolated forest patches host a wide diversity of chameleons, and many patches contain their own unique species. The actual number of chameleons in the forest biome is still unknown.

Albany thicket

Albany thicket is a small biome found in southeastern South Africa, and is sometimes classified as part of the savanna biome. It is characterised by a low, thick canopy of evergreen shrubs, stem-succulents, vines and thorny bushes. Aloes, tree euphorbias and spekboom are familiar components of the thicket. Some elements of the forest are also present in thicket areas, so it is sometimes considered a type of transition between forest and other biomes. In fact, thicket is in contact with every other southern African biome except desert and succulent Karoo. At least one species, the Eastern Cape Dwarf Chameleon, is found in thicket, although it is not restricted to this biome.

Indian Ocean coastal belt

The coastal belt is a complex mosaic of subtropical forests that are sandwiched between the sea and the savanna biome running the length of the east coast. This belt is sometimes classified as part of the savanna biome. The environment is warmer and wetter than temperate forests. There may be as many as five species of chameleon within the Indian Ocean coastal belt.

Savanna

Savanna is the largest biome in southern Africa, and is found across northern Namibia, most of Botswana and Zimbabwe, parts of Mozambique and parts of northern South Africa. Savanna also extends along the east coast into KwaZulu-Natal and the Eastern Cape. Although savanna is dominated by grasses, the biome receives enough rainfall for some trees and shrubs to grow. Savanna can be further subdivided into different types (e.g. miombo woodland) according to the tree species composition. The chameleon species most likely to be found in savanna is the Common Flap-necked Chameleon.

Grassland

The grassland biome occurs in a large swathe across the high interior plateau of South Africa, although it also extends into lower altitudes of KwaZulu-Natal and the Eastern Cape. As the name implies, grasses are dominant, with trees and shrubs being scarce or absent. Grassland is a fire-prone vegetation type requiring frequent burns to be maintained. Only a few chameleons are found in the grassland biome, most often occurring in areas where grasslands meet other biomes.

Desert

The desert biome in southern Africa is restricted to a band along the west coast, running from northern Namaqualand to Angola. This desert, the Namib, is the oldest desert on Earth. Its vast 'seas' of shifting sand began forming around 35 million years ago. A true desert, the region is very dry, receiving only about 10 mm of rain per year, yet an abundance of life forms have adapted to these conditions, and many specialised animals thrive here.

Walter Knirr/IOA

Life in the desert receives some relief from the heat and drought through frequent fogs that form along the coast as the warm air condenses over the cold ocean. Only one species of chameleon, the Namaqua Chameleon, is found in the Namib.

EVOLUTION AND THE FOSSIL RECORD

Many people do not realise that chameleons are actually lizards, in the same way that geckos and iguanas, for example, are lizards. Together with snakes, lizards form one of the major reptile groups, or orders, the **Squamata**. The other groups of reptiles are the crocodilians (crocodiles, alligators and caymans), the chelonians (tortoises, turtles and terrapins) and the tuatara (the so-called 'living fossil' of the reptiles) from New Zealand. The closest relatives of chameleons are the agamids or 'dragons', a group of lizards found in Asia, Australia and Africa. Chameleons are also related to the iguanas and their allies (Iguanids), a familiar group that occurs in the New World and Madagascar.

The closest relatives to chameleons are found in the family Agamidae. A familiar example is the Southern Rock Agama (*Agama atra*).

The relationships between chameleons, iguanids and agamids are ancient. Fossil evidence suggests they probably arose from a common ancestor more than 100 million years ago, at a time when dinosaurs still ruled the Earth. However, the oldest known fossil chameleon, *Chamaeleo caroliquarti,* is only about 26 million years old, and was found in Europe. Younger fossils have been found in Europe, Africa and even in the Far East. The distribution of fossils suggests that chameleons were previously more widespread than they are today. The only known fossils from southern Africa are 5–6 million-year-old jaw fragments from the west coast fossil bed near Langebaan.

The dwarf chameleons are all closely related, and arose from a common ancestor. Because the youngest dwarf chameleon fossil is 5–6 million years old, the common ancestor, by default, must be older. It is likely that, in one form or another, species of *Bradypodion* have been around for at least 10–15 million years, with many new species arising in the last 3–6 million years, and possibly others becoming extinct in that same period. Some species of dwarf chameleons are more closely related than others, as has been shown by genetic studies. The geographic proximity of populations plays a role, with the closest relatives usually found living near to one another.

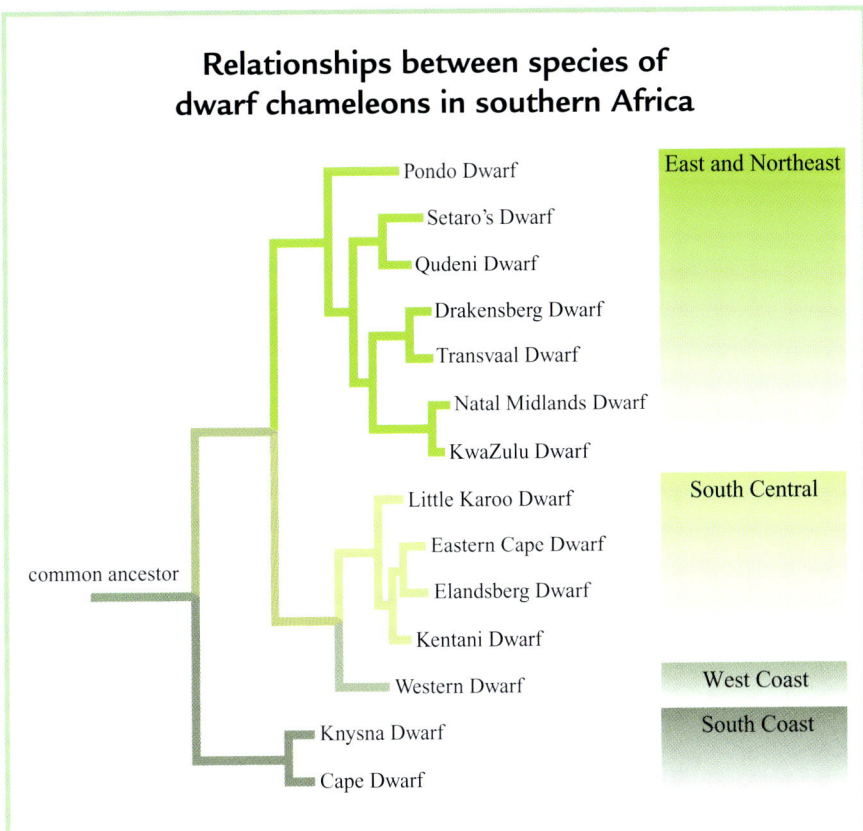

Relationships between species of dwarf chameleons in southern Africa

The above 'tree' of relationships between species of dwarf chameleons shows that the closest relatives occur in the same geographic region. In this chameleon family tree, the closest relatives also have the shortest adjoining 'branches'. For example, the Kentani Dwarf, Elandsberg Dwarf and Eastern Cape Dwarf are all closely related (they all have short branches leading to each other) and they all occur in the Eastern Cape. On the other hand, the Knysna Dwarf and Transvaal Dwarf are only distant relatives (as shown by their separated branches) and are in different geographic regions (the Knysna Dwarf occurs on the south coast and the Transvaal Dwarf in Limpopo and Mpumalanga).

Today, Madagascar boasts the lion's share of chameleons, with almost half of all species being endemic to this island, despite a total land area that is less than half the size of South Africa. Chameleons probably originated on Madagascar and, although the date of origin is not known with certainty, it is presumably within the last 90 million years. They dispersed to Africa and other areas only later, evolving into the diverse forms that are found today. East Africa enjoys the second greatest chameleon diversity, with approximately 30 per cent of all known species.

There are currently 19 chameleon species recognised in southern Africa: two of the 19 currently recognised species are in the genus *Chamaeleo* (typical chameleons), two in the *Rhampholeon* (African pygmy chameleons) and 15 in the *Bradypodion* (dwarf chameleons). Several undescribed species still require formal naming and description in the scientific literature. For the purposes of this book, we have included another six recognisable **taxa** which may prove to be either new species or just variants of known species. Therefore, there are potentially up to 25 chameleon species in southern Africa.

Each species of dwarf chameleon has a restricted distribution, and you will not find two different species in the same locality. Some species of *Bradypodion* can be found in the same area as *Chamaeleo* though, especially in Limpopo, Mpumalanga and KwaZulu-Natal provinces of South Africa. The two species of *Chamaeleo* may overlap in distribution in parts of eastern Namibia, but otherwise they are not found together.

New World 'chameleons'

Anole lizards such as *Anolis lividus* (left), which are found in the Americas, are often likened to chameleons. Although these lizards can change colour, that is really where the similarity to chameleons ends. What's more, the degree of colour change that anoles achieve is extremely limited compared to that of chameleons. However, this superficial similarity has given anoles the name 'chameleons of the New World' and led to the misconception that chameleons occur in the Americas. The comparison is misleading because, in reality, chameleons and anoles are in two completely different families: while both are lizards, they have evolved independently.

ECOLOGY AND ANATOMY

Most chameleons are **arboreal**, living in trees and bushes. This arboreal existence has resulted in a unique set of behavioural and morphological adaptations that make them unlike any other lizards. Firstly, chameleons are '**laterally compressed**', looking as if their sides have been pushed in, and giving them their thin appearance. This allows them to walk along a stick or branch relatively invisible from above or below. They can quickly reorient themselves, swivelling their body around a branch to avoid detection by predators. Some chameleons are terrestrial and have special adaptations, such as dull coloration, that allow them to blend in with the forest floor. These ground dwellers generally have stubby tails and are quite small, giving them a leaf-shaped appearance.

S van Noort

Flat-headed Pygmy Chameleon (*Rhampholeon platyceps*) blending in with its background.

The skin of chameleons is granular. The **granules** are actually modified **scales** that resemble small dots scattered over the body, spaced widely apart in places and revealing the skin beneath. Some of these granules are enlarged to form **tubercles**, which can be of a variety of different colours, often contrasting with the underlying skin.

Close-up of skin: chameleons are covered in modified scales called granules and tubercles, which can be very colourful.

Chameleons shed their skin as they grow.

As in all reptiles, chameleons periodically shed their skin as they grow. Generally, chameleons shed every few months, although juveniles grow faster than adults and must shed more often. Reptiles have 'indeterminant' growth, which means that they continue to grow throughout their entire lifetime. This is different from mammals where growth is determinant, and a maximum size is reached around adulthood. Although reptiles continue to grow, adults do so at an ever-slowing rate, requiring them to shed their skin only infrequently. Because of this type of growth, very large individuals are generally also very old.

This Eastern Cape Dwarf Chameleon changed from light to dark in a matter of seconds as a reaction to being woken by the photographer.

Chameleons are well known for their ability to change colour. These colour changes originate in special cells under the **epidermis** called **chromatophores** and **melanophores**. Both are specialised skin cells that contain pigments (colours) within them. The re-arrangement of **pigments** under the skin is controlled by nerves, and that is what brings about the colour change. Chameleons do not consciously choose to turn a particular colour; rather it is a physiological response to external conditions. Stressed or sick chameleons will turn dark, while aggressive chameleons may display highly contrasting dark and light colour patterns. Males often display vivid colours when courting females.

Chameleons are well known for their ability to catch prey with their projectile tongue, which acts like a suction cup.

Plain or paisley?

The common belief that chameleons change colour to match their surroundings for camouflage purposes is somewhat mistaken. Chameleons do not change colour to match any background, rather they have natural coloration that allows them to blend in with the background for camouflage. But there are a number of reasons why chameleons do change colour, including **thermoregulation** (becoming darker or lighter in order to warm up or cool down, respectively), and in response to behavioural stimuli. Each species has a limited range of colours and simple patterns that it can display – and these do not include turning checkered or paisley. Some species are more colourful and vibrant than others and this often depends on the type of habitat in which they live. Arboreal chameleons in closed-canopy habitats, such as forests, tend to be more brightly coloured, with a larger range of hues; this makes them more visible to each other in low-light situations. Terrestrial forest chameleons tend to be darker and fairly plain in order to match the forest floor. Chameleons in open-canopy habitats, such as fynbos or grassland, tend to be dull in colour with a limited range of possibilities for change.

A confusing situation for a chameleon?

Another iconic characteristic of the chameleon is its long, projectile tongue, which can be as long as, or even longer than the body. The tongue is controlled by a special set of bones and muscles, so that it is 'spring-loaded' at the back of the mouth. The end of the tongue is not coated with a sticky substance, but is moist and somewhat thickened. When it hits the prey, a negative pressure is created, so that it acts like a wet suction cup, helping to adhere the prey to the tongue. If the tongue hits home, prey has little chance of escaping, as the speed at which a chameleon shoots its tongue has been clocked at over 20 km/hour. Their aim is not always perfect, however, and many chameleons have to take several shots before hitting the target.

Chameleons lack molars, canines and incisors, and have only very small, rounded teeth; they are said to have 'acrodont dentition'. These tiny teeth are not good for biting, sawing, chomping or chewing prey. Instead, chameleons crush prey between their jaws and then swallow it whole. Primarily **insectivorous**, they prey upon flies, grasshoppers, crickets, insect larvae and other small **invertebrates**. Large chameleons, such as species of *Furcifer* and *Calumma*, are known to eat small mammals and birds and, occasionally, even other chameleons. The Namaqua Chameleon frequently preys upon lizards. Besides eating, chameleons must also drink regularly. They usually lick up dew and raindrops from leaves, but they can also drink from small pools of water.

S van Noort

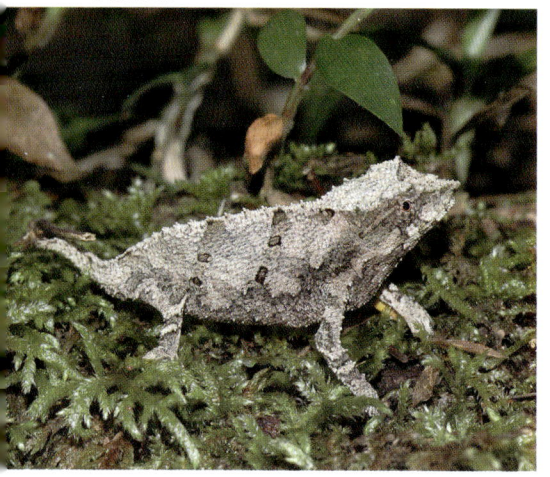

Chameleons have other interesting **anatomical** characteristics, including a **prehensile** tail that effectively acts as a fifth limb, which they use to cling to leaves and branches. Their tail is so strong that some chameleons can actually pull themselves up onto a branch using only their tail. Leaf and pygmy chameleons (*Brookesia, Rhampholeon* and *Rieppeleon*) have shortened tails that are generally not prehensile, but these chameleons are mostly terrestrial, inhabiting leaf-litter on the forest floor rather than hanging around in trees.

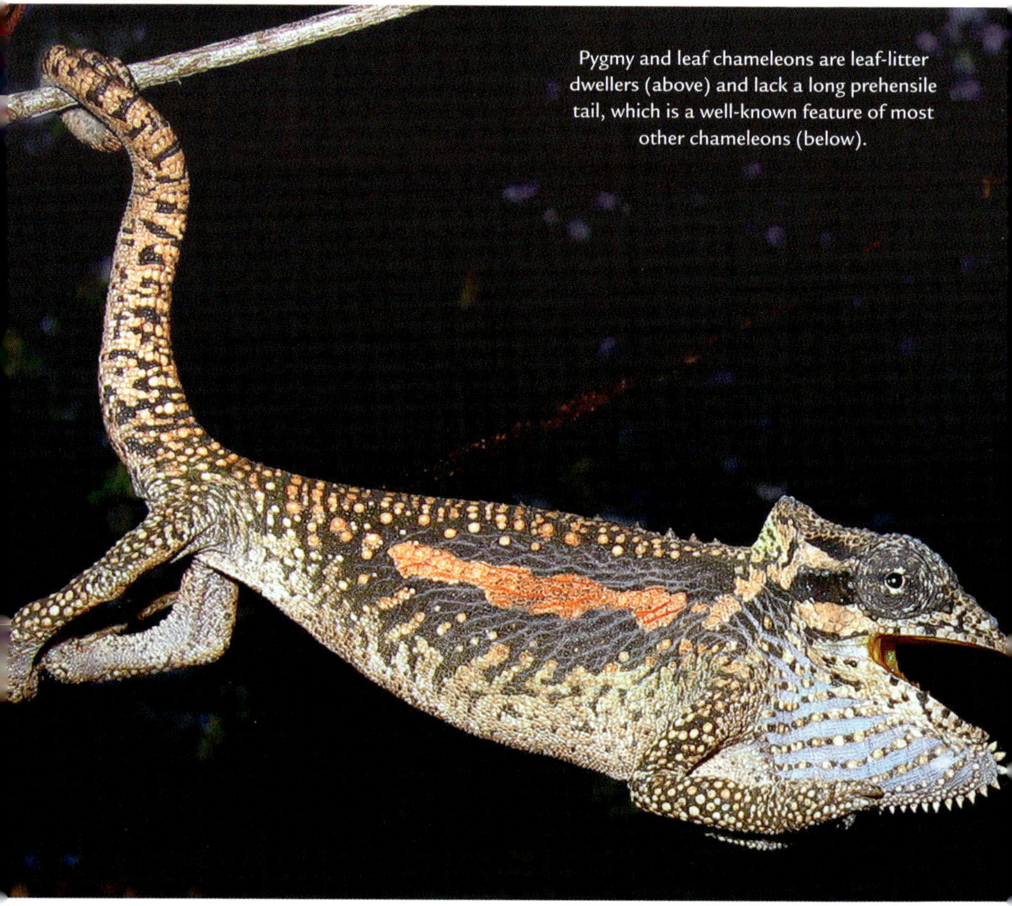

Pygmy and leaf chameleons are leaf-litter dwellers (above) and lack a long prehensile tail, which is a well-known feature of most other chameleons (below).

D Stuart-Fox & A Moussalli

Unlike most other lizards, chameleons do not lose their tail as a method of escape from predators, and they cannot re-grow it if it is damaged. Chameleons with maimed tails have been observed in the wild, suggesting that they can survive despite such a handicap. Their feet are also unique in that the **digits** are fused. On the front feet, the two outer and three inner digits are fused, while it is the reverse with the hind feet, where the three outer and two inner digits are fused. This gives their feet something of a 'claw' shape, which equips them for an arboreal existence, allowing them to grasp branches firmly without slipping.

Chameleons' most developed sense is that of sight. They have particularly sensitive retinas, highly developed optical nerves and an enlarged optical lobe of the brain. Their specially designed eye sockets allow their eyes to swivel independently, giving them a nearly 360° view of their surroundings, which means they can see in almost all directions (except directly behind the head) without moving or turning their body. How a chameleon can process all this independent information to form a picture in their brain is still a mystery to science. This ability to remain absolutely still while surveying their surroundings increases their capacity for stealth. Chameleons do not have a good sense of hearing as they lack both external ear openings and eardrums.

The sex of chameleons can be determined in mature individuals by the presence of a **hemipenal** bulge in males and the lack of this bulge in females. The bulge that can be seen on the underside of the tail at the base, **posterior** to the **cloaca**, contains the **hemipenes**.

Chameleons can swivel their eyes independently in nearly any direction. Their feet are highly modified to form a kind of claw for grasping twigs and leaves.

BEHAVIOUR AND REPRODUCTION

While much about the anatomy and physiology can be generalised for chameleons, the same cannot be said about their behaviour. With nearly 160 species, there are bound to be many differences, and the behavioural traits discussed here pertain to southern African species, mainly dwarf chameleons.

This Cape Dwarf Chameleon shows that a firm grip is required when perching for the night.

Chameleons have several interesting habits that increase their ability to blend into the background. Their movements through the vegetation are generally slow and stealthy. They creep low along branches, with hesitant steps, or with a slow, swaying jerk, which presumably mimics the movement of leaves. This way of moving is not unique in the animal world: for example, mantis insects move in a similar way as they creep along branches and vines, positioning themselves to lie in wait for prey.

While chameleons are proficient climbers, they can, and often do, move across the ground, and some species spend most of their lives on the ground. During the day, arboreal chameleons usually forage under the cover of vegetation, but towards dusk they find a suitable perch to sleep until daylight. They use their 'claw-like' feet to gain a sure grip on the branch, where they can remain swaying in strong winds and rain. It is not uncommon to find chameleons perched on the tips of thin and spindly branches, a strategy that may keep them out of the reach of some snakes. Should a snake set its sights on a sleeping chameleon, the chameleon will become aware of the vibrations made by the approaching predator, and escape by letting go and falling to the ground. During the day, they avoid predators (usually snakes and birds) by blending in with their surroundings, by flattening themselves and swivelling to the far side of branches to hide, or by dropping from their perch to the ground.

Individual chameleons show little tolerance for each other. They either avoid one another (above), or engage in vicious fighting (below).

Chameleons are sometimes called 'social', implying that they live in stable groups, form a community and have complex and long-term associations with other individuals in the group. Certain species are found in dense aggregations or in pairs (especially if the female is receptive to mating); and mate-guarding has even been observed, such as in the Mediterranean Chameleon of Europe, suggesting that at least some species of chameleons have complex social interactions at some level. Despite this, the tendency to associate with other chameleons has only been documented for a few species, and cannot be used as a general rule. In fact, chameleons can be intolerant and aggressive towards each other. True 'social' behaviour in the form of long-term associations and stable groups has not yet been established, even among dense aggregations of chameleons. In documented cases of mate-guarding, the male abandons the female when she is no longer receptive, demonstrating that their relationship is transient rather than long-term. Thus, the bulk of the evidence suggests that, for the most part, chameleons are not truly social.

None of the species in southern Africa are found in aggregations, and cannot even loosely be termed 'social'. Densities are higher in some places than others and, sometimes, more than one adult can be found sharing the same bush or tree. This is most likely the result of limited suitable habitat, either in natural areas or in urban gardens and on road verges. In high densities, they tend to ignore each other, perhaps giving the impression that they live together in peaceful harmony.

When face-offs occur, as they inevitably do, chameleons will fight, bite, threaten and chase each other. Dwarf chameleon females are particularly aggressive toward males, and usually succeed in intimidating them into retreating. In such encounters, the female will quickly turn an intense, contrasting pattern of dark and light colours, open her mouth wide and hiss aggressively, showing the bright orange interior of her mouth. This is usually enough to cause the male to beat a hasty retreat. If he hesitates, she will attack. A receptive female, however, will suspend such aggression and allow a male to approach and mate with her. She shows her willingness to mate by behaving calmly and displaying brighter colours than when she is **antagonistic**.

Top left: An agitated Western Dwarf Chameleon shows its bright gular grooves.
Top right: Aggressive dwarf chameleons hiss and show their bright orange mouth.
Above: Jackson's Chameleons often put their horns to use in brutal fights.

Territorial chameleons?

There are numerous reports of individual dwarf chameleons inhabiting urban gardens and plant nurseries for extended periods of time. These observations appear to be accurate, and preliminary studies on marked dwarf chameleons in the wild suggest that these chameleons may utilise the same area, even the same bush, day after day, at least in the short term. While this confirms that dwarf chameleons have distinct habitat preferences in the short term, it does not confirm that they actually set up and defend a territory over the long term, and the size of the home range over the lifetime of an individual has not been investigated.

In situations where chameleons appear to inhabit the same territory over a significant period of time, it may be that their natural range has been reduced because their natural habitat is fragmented by features such as manicured lawns, security walls, parking lots, roads, golf courses, shopping malls or agricultural fields. It is possible that, rather than being territorial, they have a restricted range because they are trapped inside a patch of good habitat.

Encounters between males are often aggressive, but the intensity of the conflict varies. Males size each other up, and the encounter results in either an all-out fight, after which the loser flees, or a half-hearted fight where one participant gives up rather quickly and retreats. The winner stands his ground and shows bright colours, whereas the loser generally goes dark as he makes his getaway by either hurrying away or dropping off the branch.

Mating behaviour of different species varies, but a few generalisations can be made, especially among dwarf chameleons. In this group of chameleons, males display to females to entice them to mate. The display usually takes the form of rapid head wagging and the expression of bright coloration to attract the female. If she is not interested, as is often the case, she shows aggression towards the male until he retreats. If she is interested, she allows

A mating pair of Zululand Dwarf Chameleons from Weza.

him to approach. He then mounts her from behind and twists his body and tail around her in order for his **hemipenis** to enter her **cloaca**. Chameleon mating lasts, on average, from about 10–30 minutes. Although males have two hemipenes, only one is used during each mating. Forced copulations are not common, but have been observed in dwarf chameleons and are more successful if the male is larger than the female.

Different chameleon species have different methods of reproduction: some species lay eggs (**oviparous**), while others give birth to live babies (**viviparous**). Most chameleons are oviparous, with only about 30 species being viviparous. Both species of typical chameleons (*Chamaeleo*) in southern Africa are egg layers. The female digs a hole in the ground into which she lays a clutch of some 20–60 eggs, each about 1–2 cm across. Clutch size differs between and within species, with large and old females usually laying more eggs. About 10 months later, the babies hatch and dig their way out of the nest cavity, an exhausting process that can take more than a day.

Dwarf chameleons (*Bradypodion*) are all live-bearers, as are about 14 species in the *Chamaeleo* (*Trioceros*) group. The fertilised eggs develop for several months, each inside a separate sac within the female. In due course, the female deposits the sacs on the surrounding vegetation and the babies break through their sacs by moving and stretching. Dwarf chameleons generally produce from 5–15 live babies at a time, and are capable of having several litters of babies each year. In dwarf chameleons, several males may mate with a single female, and it appears that the female stores the sperm from at least one, but perhaps all of the males, until she has eggs ready for fertilisation. The result could be that the babies in a single clutch may have several different fathers, although this has never been confirmed.

Chameleons do not show any parental care, and baby chameleons are immediately on their own. Reports of 'chameleon families' are hearsay, and are probably just chance occurrences of larger adults sharing favourable habitat with babies. Several babies are sometimes found on the same bush night after night, but this seems to be because the babies are slow to disperse, not chancing any movements that might make them detectable to predators. Eventually, the babies will disappear into the surrounding greenery. Dwarf chameleon babies take about nine months to mature, while typical

D Stuart-Fox & A Moussalli

A Zululand Dwarf Chameleon with her babies, to which she gave live birth. Baby chameleons often remain gathered together on the same bush for a while after birth, but will eventually disperse.

chameleons take about a year to reach maturity. Chameleons are generally short lived, probably not exceeding much more than five years in the wild. Large species may live longer, possibly more than 10 years in captivity.

A variety of animals prey upon chameleons, and snakes are among their main foes. The boomslang is especially fond of chameleons, as are vine snakes. Among birds, both shrikes and starlings are known to take chameleons. Redwing Starlings are often observed thrashing a chameleon against a tree or the ground, and Fiscal Shrikes are known to impale chameleons on thorns. Incredibly, small or baby chameleons are on the menu for some spiders and large predatory insects. In urban areas, domestic cats are probably responsible for the greatest predation on chameleons. This problem may be much larger than is generally realised – even the sweetest tabby is capable of being a chameleon assassin.

Chameleons have many predators, including mammals, birds, snakes, spiders and even other chameleons. Pictured here are victims of a spider (left), boomslang (below left) and Crowned Hornbill (below).

THE AMATEUR NATURALIST

HOW TO FIND CHAMELEONS

Searching for chameleons is an exciting challenge. They can be scarce and fairly patchy in occurrence in southern Africa, but with some persistence you will usually succeed in spotting a specimen or two. They are difficult to see by day because of their ability to blend in with the vegetation. The easiest way to find chameleons is by searching at night with a light. Sleeping chameleons tend to perch towards the ends of branches. Systematic scanning of bushes and trees with a strong torch or spotlight is likely to pay off. Chameleons lose their bright colours while sleeping, and as a result, their pale coloration stands out against the darker vegetation. (Because they are asleep and their eyes are closed, eye-shine will not betray their presence.) One does not need to investigate each branch in detail. Simply use a wide, steady scan, keep in mind the familiar chameleon shape, and look for its pale silhouette. Moths, spiders, bladder grasshoppers and flower seed-heads will undoubtedly fool you on occasion but, with persistence, you will eventually develop a knack for spotting sleeping chameleons.

Searching in the appropriate habitat is important. The best place to find chameleons is in undisturbed natural areas such as fynbos, renosterveld, indigenous forest, savanna or thicket, but you can start by exploring urban gardens that have good natural vegetation cover. Chameleons are even known to perch on exotic vegetation if it has the same basic structure as native vegetation. Chameleons are scarce in lands transformed by agriculture or occupied by domestic livestock. However, even in such areas, it's worth searching around fence borders where the natural vegetation tends to persist.

A chameleon-friendly urban garden has abundant indigenous vegetation and uses no insecticides.

Although reptiles are generally unpopular and feared, chameleons are an exception. Eco-tourists put them high on their list of creatures to see.

PHOTOGRAPHING CHAMELEONS

Chameleons can either be a pleasure to photograph, or they can cause pure frustration. The best way to photograph them is at night when they are asleep, before they notice your presence. Eventually your movements will awaken the sleeping chameleon, but it will usually remain stationary for several minutes after waking, often with its tail curled up in a tight spiral. This is your opportunity to photograph it. Once disturbed, chameleons are difficult to catch through the lens, as they constantly orient themselves away from the photographer. They also begin to turn dark, making any colour patterns essentially imperceptible.

Chameleons can also be photographed during the daytime on natural vegetation. Alternatively, drape a black cloth in the background to get a night-time effect and show the chameleon in greater contrast.

The use of a flash is essential at night, and a slave flash will help to reduce shadows. A flash can also be used for daytime photography, especially for close-up shots. A macro lens is highly recommended for small chameleons. Manual settings work best, and you will have to experiment by trying out different exposures. For flash photography with a macro lens, a shutter speed between 125 and 200 and an f-stop ranging from about 22 to 32 usually give the best results, in spite of what your light metre might be telling you. Small digital cameras can be surprisingly good for photographing chameleons, especially if equipped with a macro function.

CONSERVATION

The biggest potential threat to chameleons is habitat transformation. The removal of natural vegetation to make way for modern development such as shopping malls, golf courses, housing developments and even agriculture causes habitat loss and fragmentation. Chameleon species have evolved in response to specific habitat conditions, and some species are highly sensitive to changes in this regard. Habitat alterations not only decrease their numbers, but may also bring about their local extinction.

The *Critically Endangered* Elandsberg Dwarf Chameleon occurs in a restricted area in the Eastern Cape. Its habitat is threatened by extensive farming of exotic pine trees.

Two southern African chameleon species are currently listed in the IUCN Red Data Book (RDB) as being threatened. They are the *Critically Endangered* Elandsberg Dwarf Chameleon, and the *Endangered* Setaro's Dwarf Chameleon. Two other species are considered *Near Threatened* (Qudeni Dwarf Chameleon and Natal Midlands Dwarf). At the time of writing this book, we are aware of several more chameleon species that are also RDB candidates. These will be assessed over the next three years in the course of the Southern African Reptile Conservation Assessment (SARCA; see *www.saherps.net*), and will be listed in a revised RDB edition in 2009.

Land transformation has, unfortunately, become a feature of the southern African landscape and, as a result, countless plants and animals have disappeared from some areas. It is possible to provide good replacement habitat for chameleons in impacted areas, and thus to maintain viable populations even in an urban setting. It is not uncommon for chameleons to inhabit gardens, especially those gardens that provide good cover and a source of food. Attracting insects with compost heaps or manure piles will provide a steady source of food. Avoiding exotic plants and encouraging native vegetation instead will help provide preferred perches and cover for chameleons.

However, providing suitable perches is not enough – it is important that chameleon-friendly gardens should be free of insecticides. Insects form the staple of the chameleon diet, and a garden with a naturally functioning ecosystem, complete with plenty of food, will be attractive to chameleons.

Suburban areas can be enhanced by encouraging neighbours to plant indigenous plants (and discouraging the use of insecticides), thus linking many gardens together to form a network of corridors for chameleons to wander. Never build garden enclosures to keep chameleons captive: thus imprisoned, they are not able to select the appropriate conditions and may be unable to thermoregulate properly, to obtain suitable moisture (chameleons drink from dew on leaves), or to find suitable mates. All these factors could influence their natural behaviour. If chameleons do not occur in your garden, it is probably because the habitat is unsuitable for them. Bear in mind that to help maintain viable populations of chameleons in the long term, a cat-free environment would be a prerequisite: domestic cats are probably the number one killer of chameleons in an urban environment.

In some provinces of South Africa it is illegal to take chameleons into captivity, to transport or to translocate and release them into a different area. These laws were made for good reasons, and ultimately serve to safeguard our unique diversity of chameleons. For the amateur naturalist and chameleon enthusiast, this means that you should not take a chameleon home as a pet or keep it captive, even in a garden enclosure. The trade of indigenous chameleons in South Africa is thus restricted, but in some countries chameleons are popular in the exotic pet trade. Tens of thousands of wild-caught chameleons are exported each year, most of them originating from Madagascar and East Africa. The majority of such chameleons will not live longer than a few months, succumbing to stress, starvation and dehydration. Far preferable to catching and keeping chameleons is to observe them in their natural environment (which may be your garden). You can photograph chameleons and even pick them up, but always leave them where you found them. The sympathetic observer may find a chameleon crossing a busy road, or in some otherwise difficult situation. In such a case, use your discretion to take the chameleon to safety nearby.

The consequences of individual chameleons being translocated can be detrimental, affecting survival, biodiversity and the gene pool. For example, capturing a chameleon along the Garden Route and releasing it into a Cape Town neighbourhood can be problematic. Should the **introduced** chameleon live long enough, it could interbreed with local chameleons, causing mixing of gene pools. This could lead to reduced survival of the resident population, and the introduced genes may result in offspring less well adapted to local conditions. It could also introduce diseases or parasites into the native populations. Chameleons are best left where they belong – in their natural distributions.

Cham-aliens

One of the biggest global conservation concerns is the **introduction** of alien species. Humans have introduced scores of **flora** and **fauna** into areas where they do not naturally occur. A few such incidences of chameleon introductions have been reported. For example, Jackson's Chameleon (*Chamaeleo* (*Trioceros*) *jacksonii*), which is native to Uganda and Tanzania, has become established in Hawaii. A population of Veiled Chameleons (*Chamaeleo calyptratus*), originally from Arabia, has become established in Florida. Likewise, the Cape Dwarf Chameleon (*Bradypodion pumilum*) appears to have been widely distributed in southern Africa. This species is endemic to the Cape Town area, but individuals have been introduced to Johannesburg, Alexander Bay, Clanwilliam and even Windhoek in Namibia.

SECTION TWO

This section is an identification guide to the 19 species of chameleon that are currently recognised in southern Africa. These species all fall within three genera – 15 in the *Bradypodion* (dwarf chameleons), and two each in the *Chamaeleo* (typical chameleons) and *Rhampholeon* (pygmy chameleons). In addition, six recognisable taxa of the *Bradypodion* genus are introduced that are still without scientific names and descriptions. They may be found to be new species, or simply variants of existing ones – their status is still under discussion. Thus there are potentially 25 species of chameleon in southern Africa.

Common Flap-necked Chameleon.

SPECIES IDENTIFICATION

The species accounts in this book are based on a variety of morphological characteristics. Size is given as the snout-vent length (SVL) plus the tail length to equal a total length (TL). The SVL is measured from the tip of the snout to the cloaca. General scalation patterns are described, as are the patterns and colours of the **gular grooves** and of the **gular scales** which run the length of the **gular crest**. Some species have a ridge of spiny tubercles that run along the backbone. This is generally referred to as the **dorsal crest**, and the shape and positioning of the dorsal crest can often be used for identification. All these characteristics are meant as a guide for identification and should not be taken as absolutes, as individuals from within a population can vary greatly from the typical. General coloration and colour patterns are also given, but chameleons can change colour within a matter of seconds, and so your diagnosis must be made with this in mind.

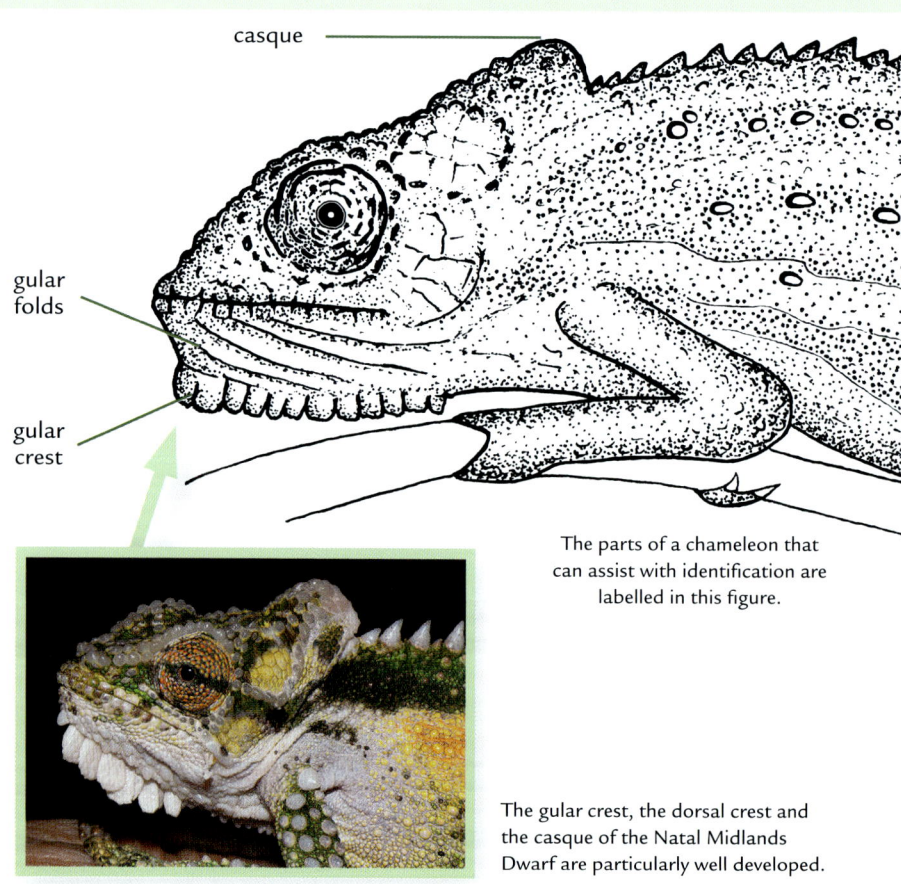

The parts of a chameleon that can assist with identification are labelled in this figure.

The gular crest, the dorsal crest and the casque of the Natal Midlands Dwarf are particularly well developed.

In this field guide, the currently accepted scientific name of each chameleon is given in its species description. This is followed by the name of the taxonomist who originally described the species and the year it was first described. Common English names for each species are also given. Some previously used common names are not especially appropriate, so alternative names are provided. The first common name that appears is the one that we recommend. This is followed by an explanation of characteristics that are helpful in identification, and a description of general coloration and patterns.

Preferred habitat and known distribution of each species are given. (As some chameleons are better studied than others, the amount and depth of information varies.) The distribution includes the 'type locality', or *terra typica*, where each specimen was first collected. Some of these chameleons were originally discovered and described more than 100 years ago, so locality names may be vague, or no longer in use. An attempt has been made to clarify the locality name if it has been changed since the original description. Distribution maps show grid squares where records exist, but chameleons should occur in the general area.

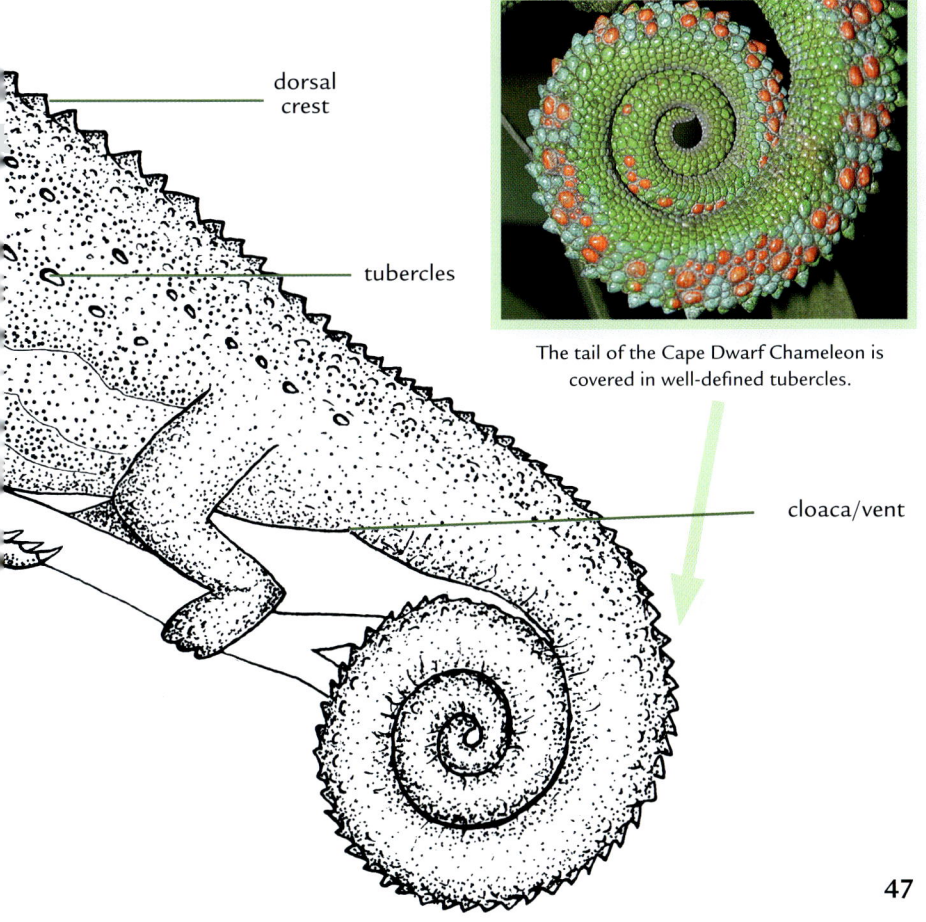

The tail of the Cape Dwarf Chameleon is covered in well-defined tubercles.

DWARF CHAMELEONS
Bradypodion

An adult dwarf chameleon in hand, showing its relatively small size.

Dwarf chameleons are endemic to South Africa. The genus is widespread, but is essentially absent from the Karoo and the Kalahari. Fifteen species are currently recognised, but several more species are currently in the process of being described. There can be a large degree of morphological variation within each species, which has flabbergasted naturalists for decades. On the other hand, different species can look similar to one another, making them hard to tell apart. Because of this, classification of species, or their taxonomy, has been difficult. A conservative approach is used in this field guide, with only the currently recognised species presented. However, there is also a section on dwarf chameleons that have not yet been described as species, but are distinct enough to deserve special mention (see pages 80–85).

In some regions of South Africa (Namaqualand, Limpopo, Mpumalanga and parts of KwaZulu-Natal), species of dwarf chameleons overlap in distribution with typical chameleons (*Chamaeleo*). Dwarf chameleons can easily be distinguished from *Chamaeleo*

All Bradypodion have a gular crest, which is especially prominent in the Knysna Dwarf Chameleon.

The Elandsberg Dwarf (above) has a low casque, whereas that of the Knysna Dwarf (opposite, bottom) is, by comparison, well developed.

because they are much smaller in size (*Bradypodion* <15 cm TL; *Chamaeleo* 20–30 cm TL) and because of the presence of a gular crest of enlarged scales, a characteristic unique to this genus. Most species have enlarged tubercles and granules on the **flanks**, whereas the granules on the body of *Chamaeleo* and *Rhampholeon* are more uniform in size. Among dwarf chameleons, identification is more difficult, but some characteristics are useful. Often, the size and shape of the **casque**, the gular crest and the dorsal crest can be helpful for identifying species.

Dwarf chameleons do not show pronounced sexual dimorphism and the result is that the sexes are difficult to distinguish at a glance. There are no clear and consistent differences between the sexes with regard to colour. In some species, sexually mature females are larger than males, but this has not been investigated for all dwarf chameleons and cannot reliably be used for distinguishing the sexes. There are also no consistent **secondary sexual characteristics** in this genus. For example, in some chameleon genera, the males have high casques and other impressive protuberances from the head, including horns. In dwarf chameleons, neither sex has such characteristics. The casque tends to be high in some species, but this is so for both sexes. However, the sex of chameleons can be determined in mature individuals by the presence of a hemipenal bulge in males and the lack of this bulge in females.

All dwarf chameleons are viviparous, giving live birth to an average of 5–15 babies at a time. The babies are fully developed when born, after a gestation time of approximately three months. They are tiny (ca 20 mm), but fully equipped for survival: they are capable of feeding on small insects and of hiding from predators, and do not require parental care. However, this is still a vulnerable time for baby chameleons, as there is a multitude of predators that can take advantage of their small size. Spiders and predatory insects are known to hunt small chameleons, and thus the pressure is on for chameleon babies to grow quickly. They become mature after about nine months. Dwarf chameleons are thought to have a life span of 3–5 years.

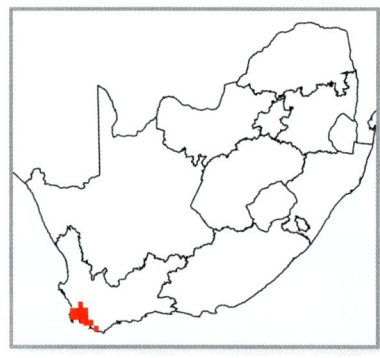

CAPE DWARF CHAMELEON
Bradypodion pumilum
(Daudin 1802)

Description One of the larger dwarf chameleons, usually reaching at least 14 cm (TL); females from Stellenbosch have been recorded up to 19 cm (TL). Individuals inhabiting the fynbos are much smaller, usually not more than 12 cm (TL). This species displays great morphological variation, possibly in correspondence with different habitat types. Three morphological forms can be recognised, but variation occurs even within these forms (see box for further details on the forms, page 52).
Coloration Varied, depending upon the morphological type (see box for further details on the forms, page 52). Generally, this species will range from greyish-green to greenish-brown to bright or even lime green. Some individuals have pink, purple, blue or orange patches on the flanks, but these patches are not obvious in the fynbos or renosterveld forms.
Habitat Found in a variety of habitats including fynbos, renosterveld, thicket, exotic and native trees and riparian vegetation. The differing morphological forms each inhabit different

vegetation types. The typical form is often found in urban gardens, in the canopy of forest patches, bushes and thicket. The renosterveld form can be found in the remnant patches of renosterveld north and west of Cape Town. The fynbos form is found in the **montane** and lowland fynbos of the Western Cape.

Distribution Southwestern corner of South Africa, from Cape Town extending as far east as the Agulhas Plain. *Terra typica:* Cape of Good Hope, Western Cape, South Africa.

Notes The typical form is unlikely to be confused with any other chameleon. The renosterveld form resembles the Western Dwarf, but tends to be more greenish and lacks the large tubercles present in the Western Dwarf. The fynbos form may be confused with the Little Karoo Dwarf, but the Cape Dwarf has wide gular lobes, rather than the thinner, pointy gular lobes typical of the Little Karoo Dwarf.

Opposite: Cape Dwarf Chameleon – typical form.
This page: Considerable colour and pattern variations occur within the typical form of the Cape Dwarf.

Different forms of the Cape Dwarf, *B. pumilum*

The 'typical form' tends to be large bodied (>14 cm TL), with the tail as long as, or longer than, the body. The casque is pronounced and colourful, especially so in larger individuals, and could be described as knob-shaped. The numerous shallow gular grooves are bright orange. The gular lobes are flattened (wider than long), with ragged edges, and are large in some individuals. One or two rows of medium-sized tubercles may be present on the flanks and regularly spaced **conical** tubercles can be present on the tail. These tubercles can be quite colourful. Overall body colour is bright green, with flank patches of blue, purple, orange or pink. Additionally, colourful granules may also be present.

The 'renosterveld form' is large bodied (>13 cm TL), with a stocky appearance. The tail is shorter than the body. There are at least two deep gular grooves with striking yellow **interstitial** skin. The gular lobes are large and flattened, with rough edges. The body is dull green-grey in colour with folds of bright yellow skin around the arms and flanks. One or two rows of medium-sized tubercles may be present on the **flanks**. The top of the head is outlined with many conspicuous spiky tubercles. Preliminary DNA results suggest this form may be a hybrid between the Cape and the Western Dwarf.

Juveniles of the Cape Dwarf tend to be small and drab. Here are two examples of the fynbos form.

The 'fynbos form' is smaller (<12 cm TL) than the typical and renosterveld forms. The tail is shorter than the body and the casque is low. The shallow gular grooves are without any notable coloration, and the gular lobes are small. Tubercles on the flanks are quite small in comparison with the typical form. The coloration tends to be greenish-brown but can be a striking green. While there are often distinct patterns on the sides, these patches lack the bright coloration of the typical form.

Three forms of the Cape Dwarf: typical form (top), renosterveld form (middle) and fynbos form (above).

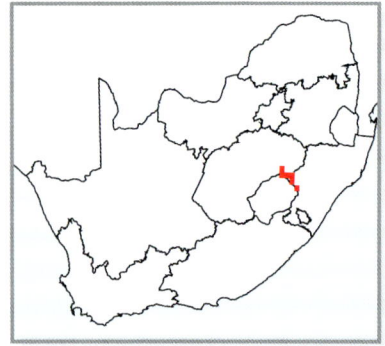

DRAKENSBERG DWARF CHAMELEON
Bradypodion dracomontanum
Raw 1976

Description A medium to large dwarf chameleon, usually not more than 14 cm (TL). The tail is about half the total length. The casque is weakly developed, and the snout short. The gular is pale and does not have deep grooves. The lobes of the gular crest are irregular in shape. The **dorsal spines** are conical, equally sized and spaced, and some may be greenish in colour. The flanks are relatively smooth, consisting of many hundreds of equally sized tiny granules. There is a series of bright, bluish-green tubercles on the flanks, and the granules on the legs can also be bluish-green.
Coloration Drab light brown, lacking distinctive coloration except for smallish, bright blue-green tubercles on the upper flanks and limbs. The cheeks and temple are often powder blue.
Habitat Mostly found on bushes in the grasslands of the Drakensberg alpine veld at altitudes of 1 500 m and above. Can sometimes be found in small remnant forest patches.
Distribution The central Drakensberg in KwaZulu-Natal and the northeastern Free State. *Terra typica*: Cathedral Peak, Drakensberg, KwaZulu-Natal, South Africa.
Notes There has been some confusion over the identity of this chameleon, and it has become obvious that there are actually two different chameleon species living in the central and southern Drakensberg. The original species description was of an individual from Cathedral Peak, and our species account is based on chameleons from that region. A second chameleon has been recorded from kloofs and alpine veld in the more southern part of the Drakensberg, near Sani Pass. It is larger and more brightly coloured with greens and yellows, and here we coin the common name 'Emerald Dwarf Chameleon' to distinguish it from the Drakensberg Dwarf. It is quite different genetically from the latter species, further suggesting that it is a different species. Thus, we treat the two separately in this field guide, and provide a separate account of the Emerald Dwarf (see section on undescribed dwarf chameleons, page 80). Records of this species from Giant's Castle have been omitted from the distribution map because their status is uncertain.

D Stuart-Fox & A Moussalli

Opposite: Home to the Drakensberg Dwarf Chameleon.
This page: Colour and pattern variation in the Drakensberg Dwarf Chameleon.

EASTERN CAPE DWARF CHAMELEON
Bradypodion ventrale
(Gray 1845)

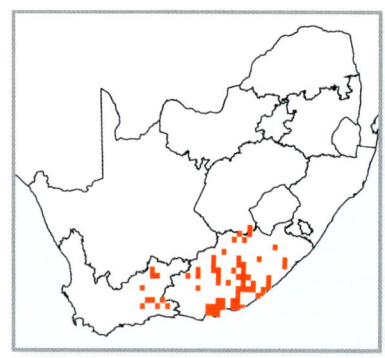

Other common name
SOUTHERN DWARF CHAMELEON

Description One of the larger dwarf chameleons, reaching over 15 cm (TL) in some cases. The tail is about 80 per cent of the body length. The casque tends to sweep back in an elongated shape, although it is not especially high. The gular region is usually pale. The gular lobes are large and fleshy, wider than they are long, and overlap towards the tip of the jaw. The dorsal crest is composed of pronounced triangular tubercles that point backwards and extend onto the tail. The body is covered in large granular scales and tubercles, giving a warty appearance. At least two rows of larger tubercles are present on the flanks.
Coloration A grey chameleon with darker grey and olive mottling, usually with a light wavy patch behind the head. There is also a light central patch on the flanks. The enlarged tubercles may be green or yellow to orange-brown.
Habitat Found in several habitat types such as thicket and Nama Karoo. Its habitat also includes the savanna/grassland mosaic of the Eastern Cape, and it has been recorded from temperate forest patches in the Eastern Cape (e.g. Amatola Mountains, Alexandria Forest). It is commonly found in the thicket that covers the lower slopes of the eastern Cape Fold Mountains, but not in the fynbos on the upper slopes of these same mountains.
Distribution This species has the largest distribution of any southern African dwarf chameleon, extending over large parts of the Eastern Cape and Great Karoo. The western limit is near Beaufort West, the northernmost locality records being Zastron and Umtata. *Terra typica*: unknown.

Eastern Cape Dwarf Chameleon – typical coloration.

Notes This species is often confused with the Little Karoo Dwarf. The latter species has gular lobes that are longer than wide, whereas those of the Eastern Cape Dwarf are wider than long. Up until recently, another species of dwarf chameleon was recognised, namely the Karoo Dwarf (*B. karrooicum*), from the drier habitats north of the Cape Fold Mountains and into the Great Karoo. The Karoo Dwarf is genetically indistinguishable from the Eastern Cape Dwarf, with only minor morphological differences. Hence, we do not treat this as a separate species.

Eastern Cape Dwarfs: Juvenile (top left); adults (top right and middle); adults displaying (above).

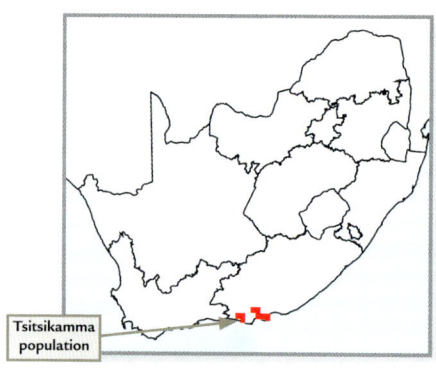

ELANDSBERG DWARF CHAMELEON
Bradypodion taeniabronchum
(Smith 1831)

Other common name
SMITH'S DWARF CHAMELEON

Description Small and slender, usually not reaching more than 11 cm (TL). The tail is approximately half the total length. The casque is low and points backwards, and the head is short. The gular is white, with 2–3 deep gular grooves that are dark maroon. These grooves are uneven in length, the topmost groove being much shorter than the others. The gular lobes are closely set and pointy. The spiny tubercles of the dorsal crest are conical, regularly spaced and extend onto the tail where they become very small.

Coloration Greenish to greenish-brown, with an obvious striped pattern along the flanks. A bright white band extends from the eye to the front leg. Another band extends from the eye to the mid-flank, where it blends with a light, metallic blue band on the central flank. This contrasts with the bands above and below, which are often yellow-red, or green-orange. The eyelids are pink or light orange.

Habitat Montane fynbos, often found perched on restios or fine-leaved vegetation such as ericas and daisies.

Distribution Restricted to the montane fynbos of the Elandsberg, but historically recorded from Schoenmakerskop on the outskirts of Port Elizabeth. Previous records from the Van Stadens Wildflower Garden could not be reconfirmed during recent surveys. *Terra typica:* near Algoa Bay, Eastern Cape, South Africa. The Tsitsikamma population is indicated by an arrow on the map.

Notes A second population of this species was previously thought to occur in the Tsitsikamma mountains but their status is still under investigation; they are genetically isolated, although morphologically similar to the Elandsberg Dwarf. Due to its small distribution, coupled with habitat loss from pine plantations, the Elandsberg Dwarf is considered to be *Critically Endangered*.

Elandsberg Dwarf – typical coloration.

Above: The Elandsberg Dwarf – typical form (top left and right); an aberrant melatistic form (middle right); two to three short, but deep gular grooves are diagnostic for this species (above left); a stressed female (above right).
Below: Chameleons from the Tsitsikamma population; their taxonomic status is uncertain.

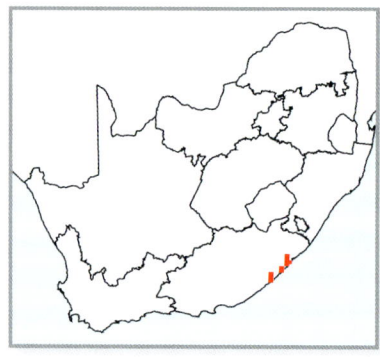

KENTANI DWARF CHAMELEON
Bradypodion kentanicum
(Hewitt 1935)

Other common name
KENTANI GRASS CHAMELEON

Description A small chameleon of just over 10 cm (TL). The tail is approximately half the total length. The casque is raised and pointed backwards, and the snout is pointy. The gular crest has small conical scales that extend onto the neck but are generally absent close to the chin. The tubercles of the dorsal crest are conical and short, widely spaced, and become very small or non-existent on the tail. A few large tubercles are scattered on the flanks.
Coloration A rather plain chameleon, brownish or olive-grey. The flanks sometimes show a pair of pale patches and scattered greenish-blue tubercles.
Habitat Prefers closed canopy habitats, occurring in trees, bushes and shrubs of the Eastern Cape coastal belt vegetation. Also known from grassland.
Distribution Known from the Kentani area and along the coast from Dwesa, north to Coffee Bay. *Terra typica:* Kentani, Eastern Cape, South Africa.
Notes Not much is known about this chameleon and it was long believed that this species occurred only in the vicinity of Kentani. Even brief surveys on the coast have produced additional localities for it, however, so it's unlikely this chameleon is as restricted as previously thought.

This page and opposite: Colour and pattern variation in the Kentani Dwarf Chameleon, a species that is found in both forest and grass habitats.

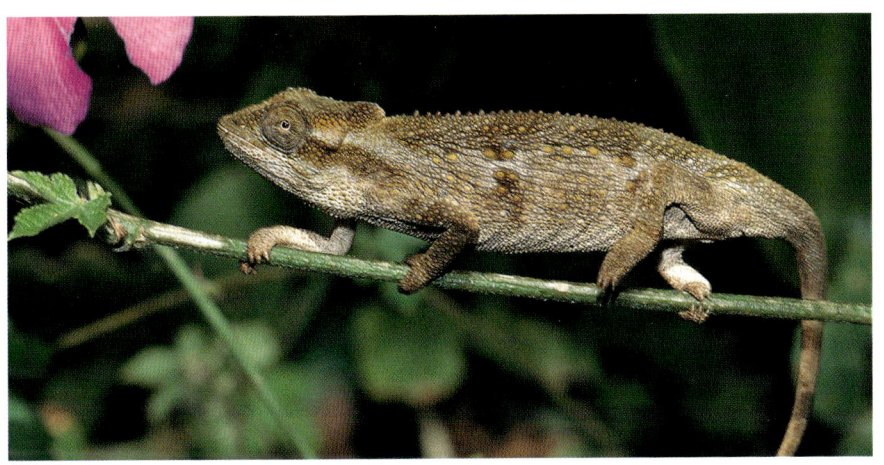

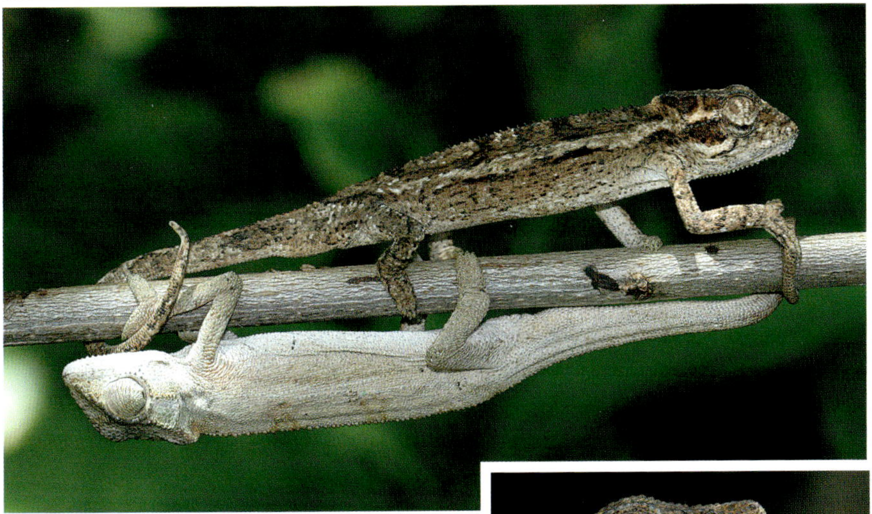

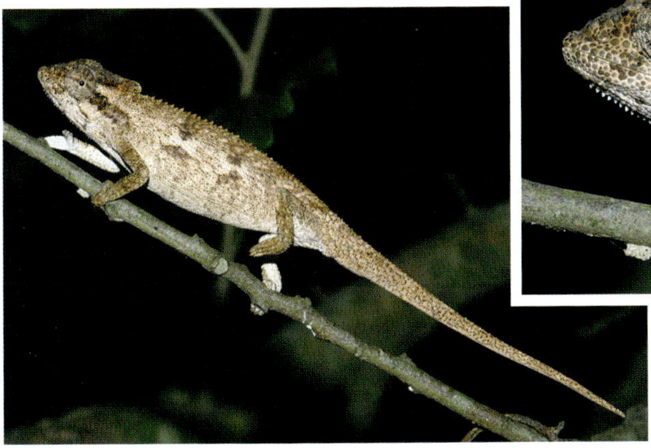

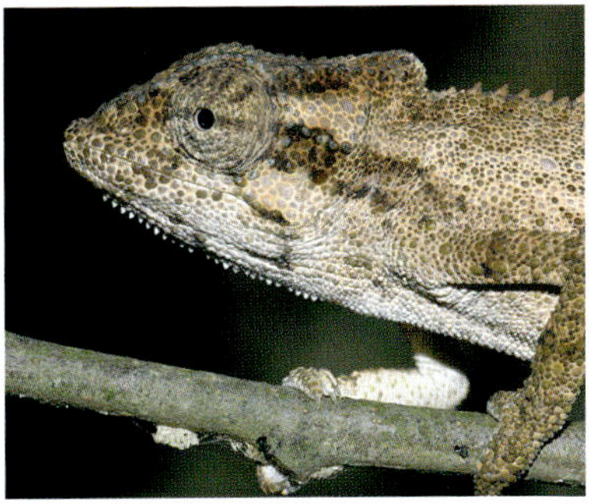

D Stuart-Fox & A Moussalli

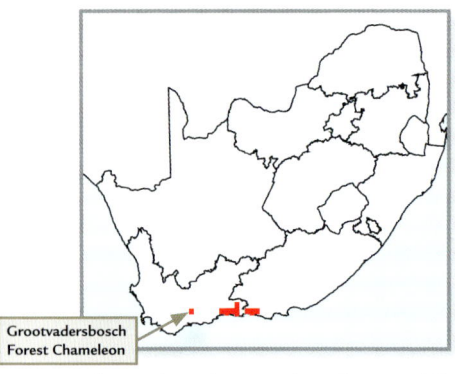

Grootvadersbosch Forest Chameleon

KNYSNA DWARF CHAMELEON
Bradypodion damaranum
(Boulenger 1887)

Description One of the larger dwarf chameleons, often reaching more than 18 cm (TL). The tail is more than half the total length. The casque is high and somewhat rounded at the tip. The lobes on the gular crest are pale in colour, flattened and flap-like with ragged edges. The gular grooves are numerous but shallow, and of a similar colour to the gular crest. The spines of the dorsal crest are conical and blunt, and extend to the end of the tail. The body is covered with unequal-sized granules, and there are usually two rows of enlarged tubercles. The tubercle arrangement of the upper row sometimes breaks into several curving rows that sweep dorsally and backwards. The smooth, pale to yellowish-orange patch of skin that is present near the armpit is diagnostic. The granules on the limbs are prominently enlarged and usually of contrasting colour to the granules on the body.

Coloration A colourful dwarf chameleon with predominantly bright green skin when displaying. Yellow and cream patches blend from bluish-violet to red on the flanks. These patches are usually large and extend from the front legs to the mid-body.

Habitat A forest species occurring in wet, coastal temperate forest in the Knysna region. Often found perched high in the canopy, but can be lower on bushes and shrubs. It also inhabits urban gardens, especially those with dense, leafy cover.

This page and opposite: Colour and pattern variation in the Knysna Dwarf Chameleon.

Distribution Restricted in distribution, occurring exclusively along the southern coast on the south-facing slopes of the Outeniqua and Tsitsikamma mountains, from Mossel Bay/George to Witelsbos. An isolated population at Grootvadersbosch is indicated by an arrow on the map. *Terra typica*: Knysna, Western Cape, South Africa.

Notes Although *B. damaranum* is the scientific name of this chameleon, the derivation of the name is a case of mistaken identity that dates back to the original species description in 1887. The name implies that the chameleon is from Damaraland, but it actually does not occur there. The taxonomist who named this species misidentified the locality in which it was found (Knysna) for Damaraland.

This chameleon is easily distinguishable from others in the region by its prominent tubercles, large casque, long tail and colourful flanks. While these same characteristics could be used to identify the Cape Dwarf, the Knysna Dwarf tends to have darker green skin with yellow/red side patches instead of lighter green skin with pink/blue/purple side patches often found in the Cape Dwarf. Both have large casques, but that of the Knysna Dwarf's tends to be more rounded and upright.

Grootvadersbosch Forest Chameleon

The Knysna Dwarf Chameleon is reported to occur, too, in the small patch of indigenous forest on the southern slopes of the Langeberg Mountains in Grootvadersbosch Nature Reserve, near Heidelberg. This isolated forest patch is surrounded by fynbos, in which the Little Karoo Dwarf (*B. gutturale*) occurs. Nonetheless, there have been several documented 'forest' chameleons found in this forest. These chameleons closely resemble the Knysna Dwarf, but genetic studies show they are quite different. It is speculated that the Grootvadersbosch Forest and the Knysna Forest were once connected, and that dwarf chameleons were free to roam this massive forest. As the climate became cooler and drier over the last several million years, the forests were fragmented and chameleons became isolated from one another. The genetic differences are probably the result of this isolation. Further studies are required to establish whether or not the Grootvadersbosch population should be regarded as a different species.

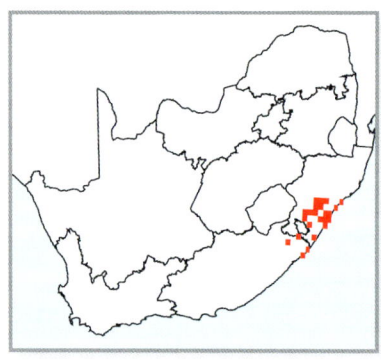

KWAZULU DWARF CHAMELEON
Bradypodion melanocephalum
(Gray 1865)

Other common names
BLACK-HEADED DWARF CHAMELEON;
DURBAN DWARF CHAMELEON

Description A small, brownish chameleon not more than 11cm (TL). The tail is as long as the body, or slightly shorter for females. The casque is low and feebly formed. The lobes of the gular crest are small and triangular, and the shallow gular grooves are pale or white. The **dorsal spines** are reduced, irregularly spaced and are weakly developed on the tail. The body is covered with equal-sized granules and there are a few scattered, flat tubercles on the flanks which may form one or two rows.
Coloration Extremely variable, ranging from uniform brown (e.g. around Durban) to greenish-brown with blotches on the flanks (e.g. around Pietermaritzburg). See box for further details on the different forms (page 66).
Habitat Known to inhabit a variety of vegetation types, including high grasses, bushes, riparian thicket, trees and urban gardens.
Distribution Found along the coast to the north of Durban, and extending south to Umtamvuna and Mkambati. Also occurs inland to Pietermaritzburg, Boston and north to Greytown. Several other populations of chameleons in KwaZulu-Natal are currently classified as the KwaZulu Dwarf, but the status of these populations is under investigation due to differences in morphology. See box for further details on these forms (page 66). *Terra typica*: 'Port Natal' (Durban), KwaZulu-Natal, South Africa.

Notes This chameleon lacks a good set of distinguishing characteristics to aid in identification. Fortunately, the nearest species to the west (the Natal Midlands Dwarf) is a large, brightly coloured chameleon with a high casque, bright white gular lobes and a tail that is longer than the body. It is easy to distinguish between adults of these two species, but juvenile Natal Midlands Dwarfs can resemble adult KwaZulu Dwarfs. To the south, the KwaZulu Dwarf borders on the distribution of the Pondo Dwarf. These species could be confused at a glance, but the casque is more developed in the latter, as are the dorsal and gular crests.

The original common name of this species, the Black-Headed Dwarf Chameleon, is inappropriate, as this chameleon does not have a black head. That name was given based on a museum specimen in which the head had turned black due to the process of preservation.

This page and opposite: Colour and pattern variation in the KwaZulu Dwarf; **gravid** female (right).

Different forms of the KwaZulu Dwarf, *B. melanocephalum*

Populations of enigmatic dwarf chameleons are known from localities scattered to the south and west of the KwaZulu Dwarf's known distribution range. These chameleons appear to differ from typical KwaZulu Dwarfs on several morphological characters, but genetic analyses have not turned up any differences between these forms. For the time being, they are treated as 'forms' of the KwaZulu Dwarf, but they may in fact turn out to be separate species. Further genetic and morphological studies are required to clarify this situation.

Form 1) Dwarf chameleons in Karkloof and Gilboa forests (north of Howick in central KwaZulu-Natal) are large-bodied chameleons with large casques, colourful spotted flanks and long tails. These chameleons are scarce, as much of their original forest habitat has been transformed into exotic tree plantations.

Form 2) Chameleons from Weza Forest, near Harding in southern KwaZulu-Natal, are similar in general appearance to the typical KwaZulu Dwarf. However, they have a series of extremely deep, highly distinctive, red throat grooves, which are lacking in the KwaZulu Dwarf. Much of the forest habitat of this area has been lost, and further surveys are needed in this area to determine the extent of occurrence and conservation status of this form.

Form 3) Dwarf chameleons from Ixopo Forest, Umzimkulu and Donnybrook areas are small chameleons, quite similar in appearance to the typical KwaZulu Dwarf, differing only in having two distinct white stripes that extend laterally along the flanks.

D Stuart-Fox & A Moussalli

This page and opposite: Different forms of the KwaZulu Dwarf – from Gilboa (opposite) and Karkloof (opposite, inset) forests; chameleons from Weza have red throat grooves (above and inset); chameleon from Ixopo reacting to a threat (below).

D Stuart-Fox & A Moussalli

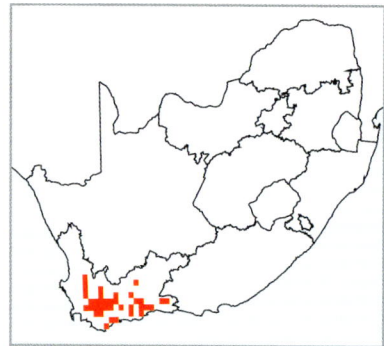

LITTLE KAROO DWARF CHAMELEON
Bradypodion gutturale
(Smith 1849)

Other common name
ROBERTSON DWARF CHAMELEON

Description A medium to large, robust dwarf chameleon which can exceed 15 cm (TL). The tail is about 70 per cent of the body length. The casque is pronounced, but not raised very high. The head is heavy-looking due to a covering of tubercles that are especially large on the crests of the temporal bones, giving it a knobbly effect. The gular consists of a series of shallow grooves, which can range in colour from yellowish to white. The gular lobes are flattened, obtusely pointed and longer than wide. They can also be conical, especially towards the throat. The spiny tubercles of the dorsal crest are large, regularly spaced and conical, and extend most of the way down the tail. Up to two rows of very large, bead-like tubercles on the flanks appear to be slightly sunken into the skin. These tubercles range in colour from black to brown, green, orange and red.
Coloration An olive-grey chameleon, although individuals can vary substantially by having patterns of green, brown and orange.
Habitat Found in the drier habitats of the Cape Floristic Region, it prefers dry fynbos, renosterveld and transition zones between fynbos and succulent Karoo. It can often be found perched on fine-leaved shrubs.
Distribution A widely distributed dwarf chameleon that's found in the Little Karoo and the surrounding mountain ranges. The westernmost records are from the southern Cederberg and it ranges eastward to the Kammanassie near Uniondale. It occurs in the western region of the Klein Swartberg, Anysberg, the Gamkaberg, Rooiberg, Langeberg and the northern slopes of the Outeniqua Mountains. It also occurs across the Agulhas Plain from Robertson towards De Hoop Nature Reserve in the southeast. *Terra typica:* unknown.
Notes This chameleon could be confused with either the Eastern Cape Dwarf or the Western Dwarf. The Little Karoo Dwarf can be distinguished by the pointed (longer than wide) or conical-shaped gular scales. The Eastern Cape and Western dwarfs have irregular-shaped gular lobes that are wider than they are long.

D Stuart-Fox & A Moussalli

This page and opposite: Colour and pattern variation in the Little Karoo Dwarf.

NATAL MIDLANDS DWARF CHAMELEON
Bradypodion thamnobates
Raw 1976

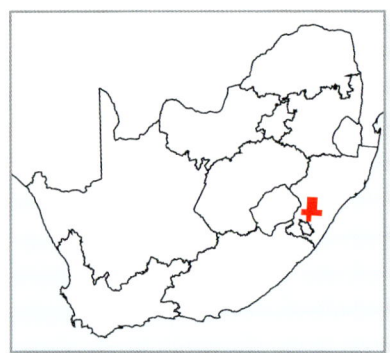

Description Large, often reaching more than 17 cm (TL). The tail is usually about half the total length in females and juveniles, but can be longer in males. The casque is very prominent, although not well developed in juveniles. Facial tubercles are prominent in adults. The gular skin and gular grooves are pure white, and the gular lobes are enlarged, fleshy and square shaped. Gular lobes often overlap each other, especially at the tip of the jaw. The dorsal spines are large, conical and sharp, often alternating with smaller spines. A few tubercles are usually present on the flanks, but the limbs tend to be covered in tubercles.
Coloration Shades of green dominate, with a patch of contrasting lighter green or yellow-orange on the flanks. The white of the gular is distinctive and extends as far back as the front legs.
Habitat Usually found in closed or thick canopy vegetation, but juveniles will enter adjacent grasslands.
Distribution The midlands of KwaZulu-Natal in the vicinity of Howick. *Terra typica*: Nottingham Road, KwaZulu-Natal, South Africa.
Notes This chameleon is distinctive due to its high casque and large, white gular lobes. It is unlikely to be confused with any other species. It is listed as *Near Threatened*.

The Natal Midlands Dwarf: adults and juvenile (inset).

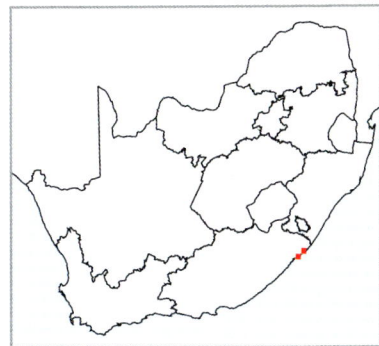

PONDO DWARF CHAMELEON
Bradypodion caffer
(Boettger 1889)

Other common name
TRANSKEI DWARF CHAMELEON

Description A medium-sized dwarf chameleon, averaging 15 cm (TL). The tail is about half the total length, but may be slightly longer than the body in males. The casque points backwards strongly, and the snout is particularly pointy. The gular is light in colour, with several pale, shallow throat grooves. The gular crest consists of fleshy lobes that are wider than long towards the chin. The lobes overlap on the chin, but become reduced in size and are non-overlapping towards the throat. The dorsal crest has large, laterally flattened spines. They are widely spaced and become smaller towards the posterior, disappearing before the tail in females, but extending about halfway down the tail in males. The first few dorsal spines may be broader than they are high. The flanks are relatively smooth, covered in small granules, sometimes with a scattering of small tubercles.

Coloration Pale brown in colour, with brown and white mottling. A dark band extends backwards from the lower half of the eye. In display, there are two distinctive emerald green, hourglass-shaped patches on the upper flanks. Coloration of the flanks gradually changes from dull red on the upper flanks to whitish to yellow on the mid- and lower flanks and tail.

Pondo Dwarf Chameleon.

Habitat A forest species, occurring in trees, bushes and shrubs of the Eastern Cape coastal belt vegetation.
Distribution Restricted in range, known only from the coast in the vicinity of Port St Johns. *Terra typica*: Pondoland, South Africa.
Notes Dwarf chameleon records from Oribi Gorge, Mkambati and Umtamvuna were thought to represent isolated populations of this species. However, genetic studies show that these chameleons are not the Pondo Dwarf, but are instead forms of the KwaZulu Dwarf.

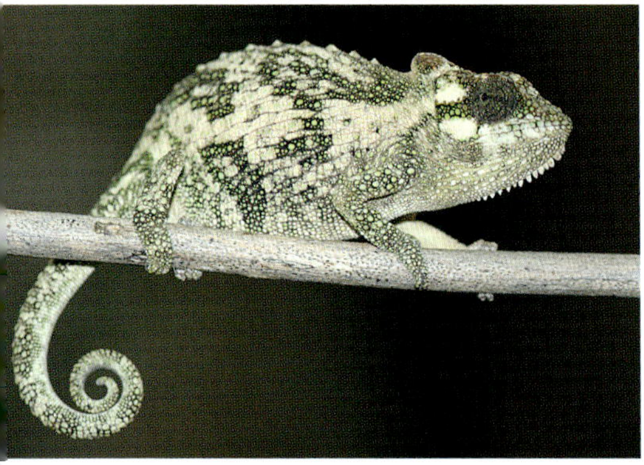

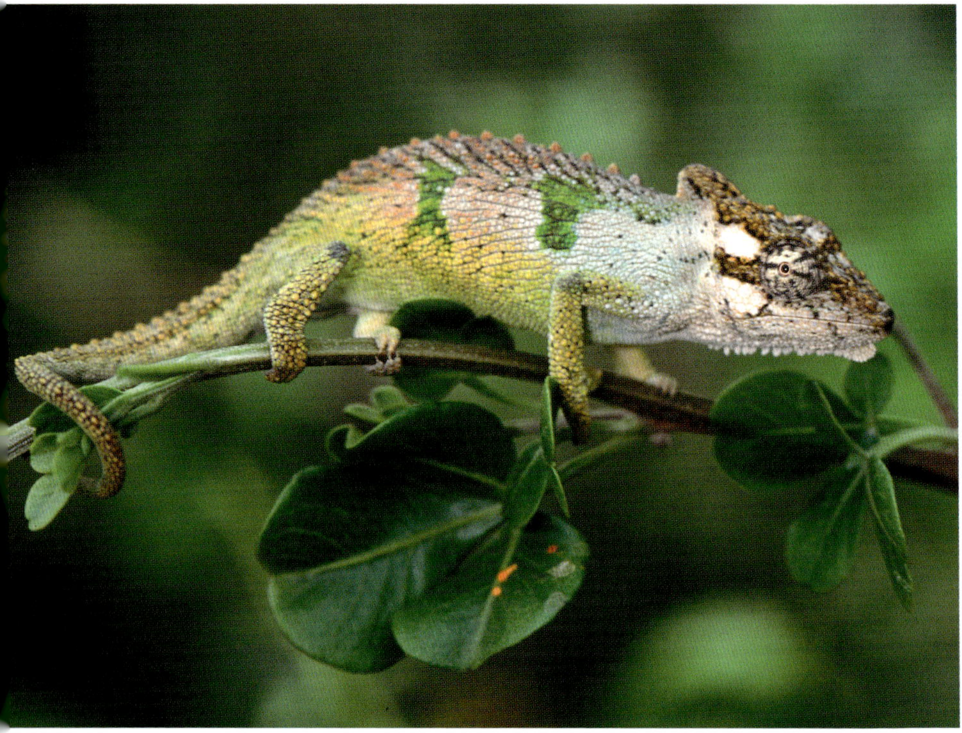

Colour and pattern variation in the Pondo Dwarf Chameleon.

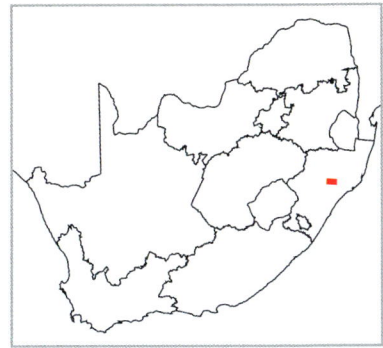

QUDENI DWARF CHAMELEON
Bradypodion nemorale
Raw 1978

Other common name
ZULULAND DWARF CHAMELEON

Description A medium-sized species, usually not longer than 14 cm (TL). The tail is about 20–25 per cent longer than the body. The head is slightly elongated, with a raised casque. The gular grooves are pale, and the gular lobes flattened. The dorsal spines are regularly spaced, and sometimes large and small spines alternate, except on the tail where they are all small. Irregularly spaced, reddish-orange tubercles extend along the sides of the tail. The flanks of the Nkandla population tend to have obvious red surface **venation** in the skin.
Coloration Orange-brown overall, with several obvious green patches on the flanks, interspersed with grey, particularly just behind the head.
Habitat Occurs in mistbelt forest in northern KwaZulu-Natal. Generally found high in the canopy, although smaller individuals have been observed on high perches in the understorey.
Distribution This species is known from only two forest patches, Qudeni Forest and Nkandla Forest. *Terra typica:* Qudeni Forest, KwaZulu-Natal, South Africa.
Notes A similar-looking species (as yet undescribed) occurs in the Dlinza, Ntumeni and Ngoya forests. Genetic studies indicate that this is a separate species. The Qudeni Dwarf is listed as *Near Threatened*.

The gular grooves of the Qudeni Dwarf are pale, but the flanks can be colourful.

D Stuart-Fox & A Moussalli

SETARO'S DWARF CHAMELEON
Bradypodion setaroi
Raw 1976

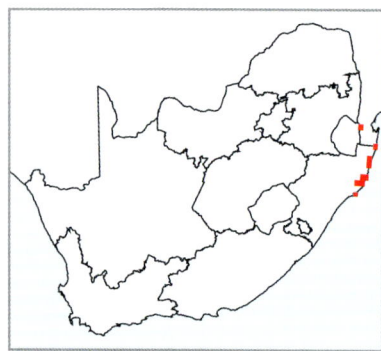

Description A medium to small chameleon, up to 12 cm (TL). The tail is about half the total length, possibly longer than the body in males, although few specimens have been examined. The casque is raised, points backwards and has a convex, sweeping shape. The head is narrow, causing the eye to look relatively large. The gular has a series of moderately deep throat grooves, with the interstitial skin a greyish-blue colour. The gular lobes are triangular in shape and irregular in size. The dorsal spines are large but widely spaced, and extend only about a third or half way down the back. Fairly large orange tubercles form a line on the flanks.
Coloration Drab brownish-grey with an irregular light orange band along the sides. In display, dark orange blotches are conspicuous on the flanks. The tail bears a series of greyish or greenish bars.
Habitat Found in trees of the coastal dune forests.
Distribution From St Lucia Estuary in coastal KwaZulu-Natal north to Lake Sibaya and just into southern Mozambique. *Terra typica*: St Lucia Estuary, KwaZulu-Natal, South Africa.
Notes Not easily confused with any other dwarf chameleon, mainly due to the distinct shape of its casque, pointy snout and the lack of dorsal spines continuing all the way down the back and tail. This species is listed as *Endangered*.

Colour and pattern variation in Setaro's Dwarf Chameleon.

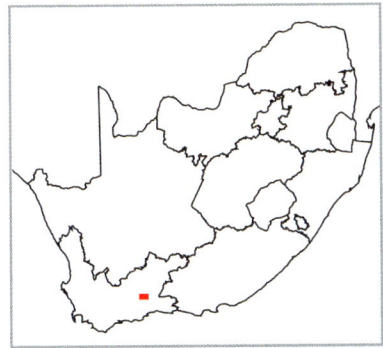

SWARTBERG DWARF CHAMELEON
Bradypodion atromontanum
Branch, Tolley & Tilbury 2006

Description A small chameleon not reaching more than 12 cm (TL). The tail is substantially shorter than the body. The casque is reduced. At least two pale bluish gular grooves are present, flanked by several shallower grooves with yellowish interstitial skin. The gular lobes are small and irregularly shaped. The dorsal spines are weak and extend onto the tail. A row of small, light green tubercles run along the flanks.
Coloration This chameleon is generally a mottled green colour. The eyelid is a pumpkin-orange colour, with a dark stripe running lengthwise across the eye. The sides of the casque tend to be black, and a dark band runs from the eye to the front leg.
Habitat Montane fynbos, often found on restios, daisies and fine-leaved bushes.
Distribution Known only from the Groot Swartberg, around the Swartberg Pass, and extending into the mountains both to the east and west of the pass. Not known from the lowlands around the Groot Swartberg.
Notes This species occurs in close geographic proximity to the Little Karoo Dwarf, and the two could be confused. The Swartberg Dwarf can be distinguished by the small, lobe-shaped (rather than conical) gular scales, and the reduced casque as compared to the Little Karoo Dwarf.

Colour and pattern variation in the Swartberg Dwarf Chameleon, and the habitat in which it is found.

WESTERN DWARF CHAMELEON
Bradypodion occidentale
(Hewitt 1935)

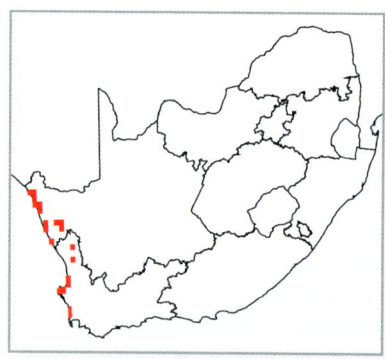

Other common name
NAMAQUA DWARF CHAMELEON

Description Large bodied, reaching up to 16 cm, with a stocky appearance. The thickened tail is about 60–70 per cent of the body length. The casque is well developed and raised. The most distinguishing characteristic is the set of 2–3 deep gular grooves that run all the way from the tip of the jaw back to where the arm joins the body. These grooves can be a variety of colours, the most notable being deep velvety purple-black, bright yellow or electric orange. The lobes of the gular crest are large and flattened, and overlap substantially, especially towards the tip of the jaw. The dorsal crest consists of large conical tubercles that extend onto the tail, where they become reduced in size. One or more rows of enlarged tubercles run along the flanks, with smaller scattered tubercles near the belly. Prominent tubercles are also present on the sides of the tail.
Coloration Generally a dull, smoky grey, with a thin, wavy lighter patch running the length of the body.

Typical coloration of the Western Dwarf Chameleon.

Colour and pattern variation within the Western Dwarf Chameleon;
in display (top left); showing the long, deep gular grooves (inset).

Habitat Typically found in undisturbed strandveld along the west coast of South Africa.
Distribution Distributed in a narrow belt along the west coast of southern Africa, from Melkbosstrand (north of Cape Town) to Lüderitz in Namibia. This chameleon has also been recorded inland, outside of the strandveld vegetation, near Kamieskroon. *Terra typica*: Namaqualand, Western Cape, South Africa.
Notes The Western Dwarf should not be confused with the Elandsberg Dwarf, another species that has several deep throat grooves. The grooves in the Elandsberg Dwarf are always deep maroon, and do not begin at the tip of the jaw. The Elandsberg Dwarf is also much smaller, and has a reduced casque and small gular lobes. Their distributions are separated by about 700 km.

The distribution of the Western Dwarf reaches its southern limit at Melkbosstrand. The Cape Dwarf has also been reported from this locality, but there are apparently no records of the two species being found together on the same bushes. They are unlikely to be confused, since the typical Cape Dwarf is much brighter in colour, is not as heavy set and has a tail that is longer than the body. Confusion with the renosterveld form of the Cape Dwarf is, however, possible. The two are distinguishable by the head shape, especially the casque, which sweeps back to a point in the Western Dwarf, but is more blunt and erect in the Cape Dwarf. The snout is more pointed in the Western Dwarf.

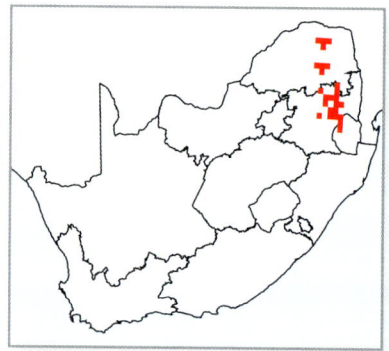

WOLKBERG DWARF CHAMELEON
Bradypodion transvaalense
(FitzSimons 1930)

Other common name
TRANSVAAL DWARF CHAMELEON

Description A large chameleon, often exceeding 15 cm (TL). The tail is at least half the total length, and up to 10 per cent longer in some individuals. The casque is well developed and points upwards. A number of shallow gular grooves are present, which are paler than the surrounding skin. The lobes of the gular crest are flattened, rectangular shaped and ragged at the bottom. The first few of these scales are usually overlapping and large. The spiny tubercles of the dorsal crest are medium to large, conical, fairly widely spaced and extend part way down the tail. The small tubercles on the flanks are usually arranged in a few rows.
Coloration Very variable depending on locality. Generally pale olive-green or pale brown, but some individuals have a striking orange hue to the skin. Coloration can be uniform, or may consist of a distinctive alternating pattern of dark brown and light-yellow lateral bands.
Habitat Prefers forested or thick, bushy habitats, usually with closed canopy. Not found in savanna areas outside forest fragments, but not uncommon in urban gardens with plenty of good cover.
Distribution Along the Drakensberg escarpment in Mpumalanga and Limpopo, with records from Wolksberg, Woodbush, Haenertsburg, Graskop, Mount Sheba, Barberton, The Downs, Pilgrims Rest, Mariepskop and Sabie. Also found in the Soutpansberg and Blouberg, north of Polokwane in Limpopo. *Terra typica*: Haenertsburg, Mpumalanga, South Africa.
Notes The taxonomy of the Wolkberg Dwarf is enigmatic, and some researchers suggest that it actually consists of up to nine species. Morphological differences, including scalation, coloration and patterning of chameleons from different localities, are the cause of this confusion. Preliminary genetic studies suggest that there are several different groups of the Wolkberg Dwarf. It is not yet known if these groups correspond to different species.

This page: Wolkberg Dwarf Chameleons: male from Woodbush in display and Magoesbaskloof male, not displaying. **Opposite:** Colour and pattern variation in the Wolkberg Dwarf.

Both from Woodbush

Soutpansberg

Barberton

Elandsvalley

Mt Sheba

Graskop

The great unknown:
UNDESCRIBED DWARF CHAMELEONS

Several species of southern African dwarf chameleons are currently still without an official scientific name. Taxonomists are trying systematically to unravel the complexities of relationships between these chameleons, and with the aid of genetic analyses, they are starting to gain a better understanding of the true chameleon richness of this region. A few noteworthy examples of possibly new dwarf chameleon species are discussed below.

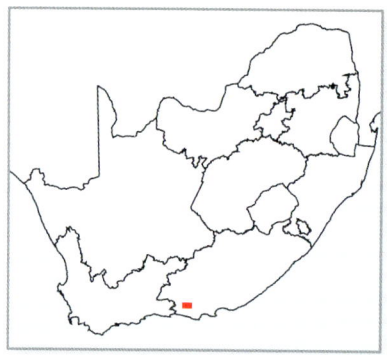

BAVIAANSKLOOF DWARF CHAMELEON

Description A small chameleon not exceeding 11 cm (TL). The tail is about 80 per cent of the body length (SVL). The casque is low and pointing backwards, and the snout is blunt. The gular crest is feebly developed. The body is covered in yellow-brown granules, while a scattering of medium-sized, dark tubercles are on the flanks. The dorsal spines are very small and irregularly spaced and become reduced toward the tail.
Coloration Basically the same as the Beardless Dwarf, i.e. pale grey or greenish in coloration with two dark patches on the flanks. The two patches are not fused in the Baviaanskloof Dwarf, but are fused in the Beardless Dwarf.
Habitat Dry montane fynbos, often found on ericas, daisies, restios or other fine-leaved shrubs.
Distribution Restricted to the Baviaanskloof Mountains. It is found neither in the Baviaanskloof valley, nor in the Kouga Mountains to the south.
Notes It was previously thought that this chameleon was the same species as that occurring in the Kouga Mountains (the Beardless Dwarf Chameleon), since there is a strong resemblance between the two. Recent genetic studies have revealed that the Baviaanskloof Dwarf Chameleon is distinct, and that these two chameleons are **cryptic species**.

Baviaanskloof Dwarf Chameleon.

BEARDLESS DWARF CHAMELEON

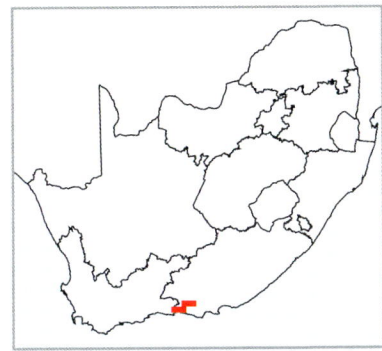

Description This is a small chameleon not exceeding 11 cm (TL). The tail is about 80 per cent of the body length. The casque is very low and not well defined. The gular is white and the gular lobes are so small as to appear nearly absent. Irregularly spaced tubercles are present on the flanks. The dorsal spines are small, conical and closely but irregularly spaced, becoming more widely spaced on the tail.

Coloration A drab chameleon that is usually greyish brown or greenish. Two dark patches on the flanks are often fused together and can be a ruddy orange or brown. The tubercles on the flanks tend to match the colours of the patches. Colours can become very pale, but the patches are almost always visible.

Habitat Dry montane fynbos, often found on ericas, daisies, restios or other fine-leaved shrubs.

Distribution Known from the Kougaberg and the northern slopes of the Tsitsikamma mountains, but not from the southern, wet, forested slopes.

Notes This chameleon is presently not a described species, but both morphological and genetic studies indicate that it is distinct.

The Beardless Dwarf Chameleon has a poorly formed gular crest.

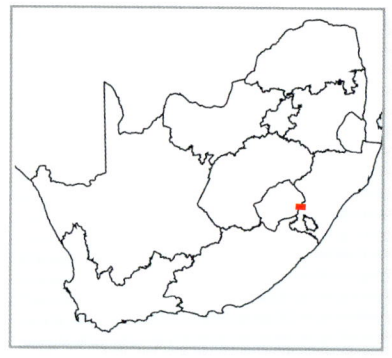

EMERALD DWARF CHAMELEON

Description A medium to large dwarf chameleon reaching 14 cm (TL). The tail is longer than the body. The casque is high and sweeps back to form a knob. The **gular** region is yellow and the interstitial skin pale. The scales of the gular crest are large and flattened, with ragged tips, and become smaller towards the back of the throat. The dorsal spines are large, irregular in shape, and very close together, overlapping in some cases. The spines become smaller as they extend onto the tail. The flanks are covered in light green granules and tubercles. Tubercles close to the dorsal spine are conical.

Coloration Shades of emerald green and lemon yellow, with dark, irregular-shaped patches scattered over the head and body. The yellow head and chest contrast with the generally green body.

Habitat Thick vegetation and alpine veld of mountain slopes.

Distribution Known only from the middle to high altitudes of the southern Drakensberg.

Notes There has been some confusion over the dwarf chameleons in the Drakensberg. The Drakensberg Dwarf Chameleon (*B. dracomontanum*) was originally described from Cathedral Peak in the central Drakensberg. That species is medium sized with a low casque, small gular scales, inconspicuous tubercles and its coloration is drab brown. On appearance alone, it is apparent that the Emerald Dwarf Chameleon is not the same as the Drakensberg Dwarf, now confirmed by genetic testing. Thus, there appear to be two species in the Drakensberg, one from the central northern Drakensberg (related to *B. transvaalense*), and one from the southern Drakensberg (related to *B. melanocephalum*). The latter chameleon has not yet been described as a species.

Emerald Dwarf Chameleon: juveniles (left) and adult (above and inset).

GROENDAL DWARF CHAMELEON

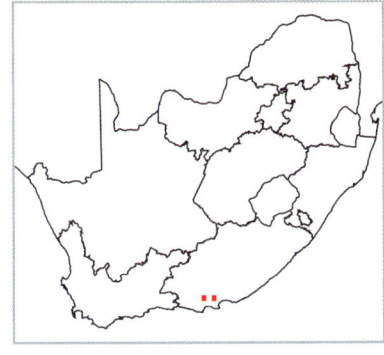

Description A small dwarf chameleon, not exceeding 11 cm (TL). The tail is about 80 per cent of the body length. The casque is very reduced, to the point of being indistinct. The snout is pointy, but short. The gular crest consists of small, non-overlapping, pointy lobes, with the first two or three on the chin being rounded and wider than the rest. The dorsal tubercles are very small and regularly spaced, extending partway down the tail. The flanks are covered in small granules, with two irregular rows of dark tubercles.

Coloration Light brownish, and generally dull in coloration. The flanks have two dark roundish blotches.

Habitat Dry montane fynbos, often found on ericas, daisies, restios, or other fine-leaved shrubs.

Distribution Known only from the Groot Winterhoek Mountains of the Eastern Cape, northwest of Port Elizabeth.

Notes This is another cryptic species that was recently discovered during genetic studies carried out on dwarf chameleons in the Eastern Cape. It resembles the Baviaanskloof Dwarf Chameleon and the Beardless Dwarf Chameleon. Not much research has yet been done on this chameleon, especially in respect of defining morphological characteristics that can set it apart from its relatives.

Colour and pattern variation in the Groendal Dwarf Chameleon.

NGOME DWARF CHAMELEON

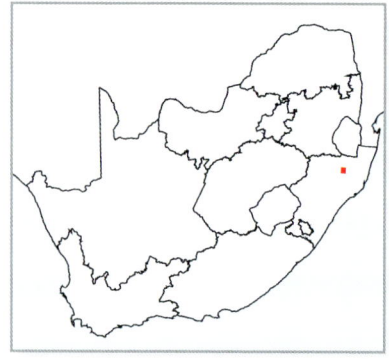

Description A medium to large chameleon, sometimes reaching more than 15 cm (TL). The tail is 10–20 per cent longer than the body. The casque is high and sweeps back. The gular tends to be the same colour as the body, but in a paler shade. The gular scales are lobe shaped and fleshy. The dorsal spines are conical, widely spaced and point somewhat backwards. At least one row of small tubercles is present on the flanks, and sometimes there is a second, lower row of smaller tubercles.
Coloration Individual coloration varies widely, from basic green to a blue tinge over the body. A patch of contrasting orange colour on the flanks is sometimes bordered by a dark outline especially when in display.
Habitat Afro-temperate forest.
Distribution This chameleon is only known from Ngome Forest, KwaZulu-Natal.
Notes Presently not described as a species, but both morphological and genetic studies indicate that it is distinct.

Ngome Dwarf: relaxed (top) and in display (above).

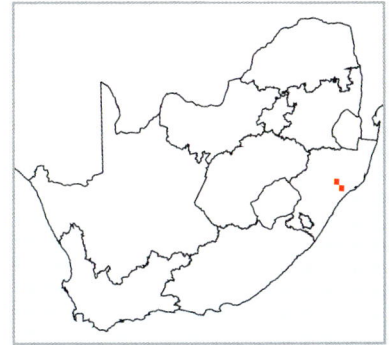

uMLALAZI DWARF CHAMELEON

Description A small to medium chameleon not usually more than 12 cm (TL). The tail can be up to 10 per cent longer than the body. The casque is raised, forming a small knob and the snout is pointy. The throat grooves are shallow and irregular, with pale blue interstitial skin. The gular lobes are white, conical in shape, somewhat irregular in size and fairly closely spaced. The dorsal spines are triangular, irregular in size, become very small and widely spaced posteriorly, and do not extend onto the tail. A network of rather deep venation on the flanks is present, showing dark red skin below. This venation is within a patch of scattered reddish-orange tubercles on the flanks. A neat row of reddish-orange tubercles runs just below the dorsal crest. These chameleons are very similar to the Qudeni Dwarf, especially when compared with the Nkandla population, but DNA studies show they are different species.
Coloration A light brownish-orange chameleon with a white belly. The brownish colour becomes paler towards the tail. The flanks are speckled with red-orange tubercles, and the flanks have a large reddish patch, within which are several smaller greenish blotches. Two indistinct stripes extending from the eye to the neck have this same greenish colour.
Habitat Prefers the canopy and high perches of coastal scarp forest.
Distribution Found in only three forests in KwaZulu-Natal: Dlinza, Ntumeni and Ngoya forests.
Notes The common name comes from the municipality where Dlinza, Ntumeni and Ngoya forests are situated.

Colour and pattern variation in the uMlalazi Dwarf Chameleon.

TYPICAL CHAMELEONS
Chamaeleo

This group of chameleons is widespread across Africa and contains more than 50 species. The genus also contains the only four chameleon species that are found outside of Africa and Madagascar. There are only two species of *Chamaeleo* in southern Africa. All species of *Chamaeleo* are medium to large, the biggest ranging from about 50–60 cm. The largest is *Chamaeleo* (*Trioceros*) *melleri* which can reach more than 60 cm. Most have strongly prehensile tails, which they use for grasping branches. Some members of this genus have horns, although the two species found in southern Africa are not horned. Most species are oviparous (lay eggs), although some species in East Africa – in the *Chamaeleo* (*Trioceros*) group – are viviparous. Both southern African species are oviparous. About 1–2 months after mating, the female digs a hole in which she lays a clutch of 20–60 eggs. The hole is usually dug with the front legs, but Common Flap-necked females have been observed using their head to excavate soil. Incubation time can vary greatly, but is generally around three months for the Namaqua and five months for the Common Flap-necked, although periods of more than a year have been reported for the latter species. The hatchlings are between 3 and 5 cm (TL), and grow quickly, reaching maturity after about a year.

The taxonomy of the genus is still in flux, and the sub-genus *Trioceros* (three-horned chameleons) is currently considered to be within the genus *Chamaeleo*. This grouping is contentious, however, and a re-assessment of the *Trioceros* will undoubtedly produce some interesting results.

In parts of South Africa (Namaqualand, Limpopo, Mpumalanga and into KwaZulu-Natal), *Chamaeleo* overlaps in distribution with some species of *Bradypodion*. *Chamaeleo* can readily be distinguished from *Bradypodion* because they are much larger (*Chamaeleo* 25–30 cm TL; *Bradypodion* <20 cm TL) and they have a more bulbous-shaped head with a somewhat convex top.

Above: Africa's largest species, Meller's Chameleon.
Right: Jackson's Three-horned Chameleon.

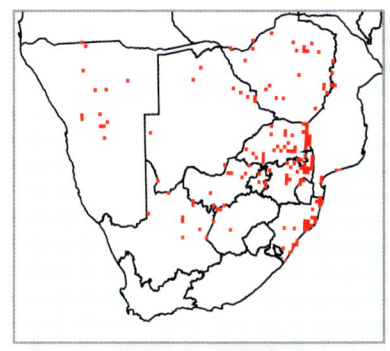

COMMON FLAP-NECKED CHAMELEON
Chamaeleo dilepis
Leach 1819

Description One of the larger chameleons in southern Africa, reaching up to 35 cm (TL). The tail is as long as or longer than the body, and is strongly prehensile. This chameleon is characterised by a skin-flap that extends from the casque over the neck. The size of this flap varies according to geographic location, tending to be smaller in the southern part of its range. The casque is raised slightly and the top of the head is concave with a centre ridge. The deep throat grooves are coloured bright orange, or sometimes yellow, contrasting against the bright green of the skin. A series of bright white triangular scales extends from the gular, past the throat and all the way down the belly to the cloaca. The tightly packed spiny tubercles of the dorsal crest are green in colour, conical in shape and regularly spaced. These spines become progressively smaller proceeding down the back, disappearing entirely towards the tail. The flanks are smooth and lack enlarged tubercles.

Coloration Usually a bright lime green and can exhibit intense patterns of dark barring or spotting when disturbed. Coloration in some individuals can be yellowish-green, brownish-green or even orange. The flanks are marked with a distinct white lateral stripe, but the size and shape of this stripe varies from one individual to the next. One or two other smaller white patches may be present above the white stripe, nearer to the shoulder. The lips are usually lined with white.

Common Flap-necked Chameleon showing bright orange gular grooves and white triangular scales that extend down the belly.

Habitat Found in a variety of habitats, including coastal forest, woodlands, thicket and savanna.
Distribution In southern Africa, this species is distributed in Mozambique, Zimbabwe, Botswana and northeastern Namibia, and in the northern areas of South Africa extending into KwaZulu-Natal. It also occurs outside southern Africa, ranging across most of tropical Africa. *Terra typica*: Gabon.
Notes This chameleon is not likely to be confused with any other in southern Africa due to the presence of a flap extending from the casque over the neck, its bright green colour, the distinct white stripe along its flanks and the bright white triangular scales extending down the belly. Baby Common Flap-necked Chameleons can be confused with small dwarf chameleons, as they tend to lack the distinguishing characteristics of the adult.

Common Flap-necked Chameleons: juveniles (right); female with mud on her forehead, indicating she has recently dug a nest (below left); an orange individual from Zambia (below right); spots appear all over the body when stressed (bottom left); a grey individual from Mozambique (bottom right).

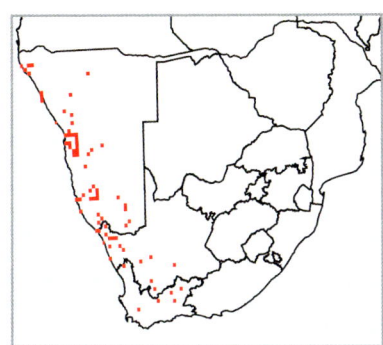

NAMAQUA CHAMELEON
Chamaeleo namaquensis
Smith 1831

Description One of the largest chameleons in southern Africa, reaching up to 25 cm (TL). The tail is much shorter than the body. The casque is large, sweeps up to a point and is crested with bony spines, as is the brow above the eyes. The gular grooves are paler than the body, often a light burnt-orange colour. The gular crest is absent. The dorsal crest consists of a series of very large, knob-shaped spikes, a character that easily distinguishes it from other southern African chameleons. These are widely spaced and become smaller towards the tail, but do not extend onto the tail. Each spike is made up of several tubercles. The flanks are smooth and lack enlarged tubercles. The legs are disproportionately long compared to other southern African chameleons.

Coloration Light brown, green or burgundy overall and often speckled with darker spots and blotches, especially when disturbed. A row of several round, white patches is usually evident on the flanks. The coloration tends to resemble camouflage gear.

Habitat Mainly terrestrial chameleon, and can be found walking across desert, sandy areas and gravel plains. It will sometimes perch in bushes.

Distribution From the southwestern Karoo of South Africa into Namaqualand, through western Namibia to southern Angola. Recorded as far south as Sutherland in the Western Cape. *Terra typica*: mouth of the Orange River, Northern Cape, South Africa.

Notes Morphologically, this chameleon is easily distinguishable from all other species in southern Africa. In extreme eastern Namibia it may overlap in distribution with its relative, the Common Flap-necked Chameleon. The two species need not, however, be confused with each other because the Namaqua Chameleon has an extremely short tail and has large, nearly square dorsal spines. It lacks the neck flaps, the distinct white stripe and the bright white gular scales that extend onto the belly of the Common Flap-necked Chameleon.

J Marais

B Phillips

Namaqua Chameleons are generally found on the ground.

89

PYGMY CHAMELEONS
Rhampholeon

Chapman's Pygmy Chameleon.

Pygmy chameleons are also known as African leaf chameleons or stump-tailed chameleons. They are small to medium in size, ranging from as little as 5 cm up to 13 cm (TL). They have a short, stubby tail that is generally not prehensile; there is no gular crest. They are usually terrestrial, inhabiting forest floors where they blend in with the leaf-litter and feed on small **invertebrates**. At dusk they ascend into the vegetation to find a safe perch for sleeping. All the species are **oviparous**, and the eggs take between 1 and 2 months to develop. The females either find a hiding place under the leaves and mosses to lay their eggs, or burrow a small nesting tunnel where they lay a clutch of 1–15 eggs. The small, oval eggs (approximately 10 mm long) hatch within a few months, producing tiny babies that are less than half a centimetre in size.

Boulenger's Pygmy Chameleon.

There are 14 species of pygmy chameleons, and just two of these occur in southern Africa (*Rhampholeon marshalli* and *R. gorongosae*). The rest occur in East, Central and West Africa, and most are limited to isolated indigenous forests. It appears that the Spectral Pygmy Chameleon (*Rhampholeon spectrum*) has the widest distribution, ranging from coastal West Africa to the interior of the continent in the Democratic Republic of Congo. However, even within this large distribution, it is found only in isolated patches of closed canopy covering.

An as yet undescribed species of pygmy chameleon from central Mozambique.

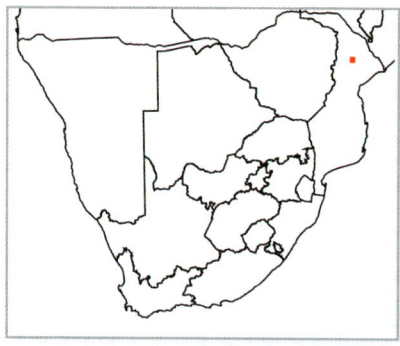

GORONGOZA PYGMY CHAMELEON
Rhampholeon gorongosae
Broadley 1971

Description A small chameleon with females attaining a maximum of just over 10 cm (TL). Sexual dimorphism is striking, with males typically not longer than 5 cm. The tail is much shorter than the body, ranging from 44–53 per cent of the SVL.

Coloration Light brown to buff (males), or light olive green (females), with no obvious distinctive patterns.

Habitat Found on the vegetation or on the ground of subtropical wet montane forests.

Distribution Restricted to Gorongoza Mountains in central Mozambique. *Terra typica*: Gorongoza Mountains, Mozambique.

Notes The male Gorongoza Pygmy Chameleon is known to ride on the back of the much larger female. This is presumably so that the male can remain with the female through the mating season.

Gorongoza Pygmy Chameleons: a male riding on the back of a female (left); male (below).

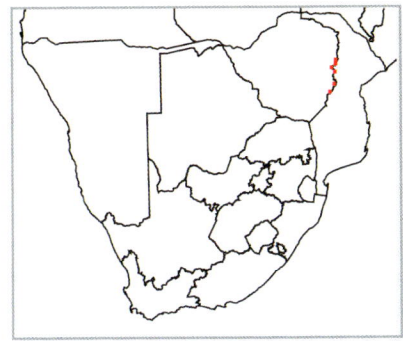

MARSHALL'S PYGMY CHAMELEON
Rhampholeon marshalli
Boulenger 1906

Description A small to medium chameleon. Females are larger than males, generally reaching up to 13 cm. The tail is much shorter than the body (50–75 per cent of SVL), but despite being short, it is slightly prehensile. There is a small bulbous, fleshy protuberance on the snout, giving the head a long shape overall. The casque is virtually absent. The gular crest is absent. The dorsal spines are very widely spaced, and do not extend onto the tail. The body is covered in small granules, with a row of tubercles extending from the front legs to the tail.

Coloration Dull brown, olive green or dark orange with indistinct darker patches.

Habitat Prefers high-elevation, humid forests. Inhabits the forest-floor leaf-litter and also clambers onto vegetation.

Distribution Montane forests in eastern Zimbabwe, and just within central Mozambique on Snuta Mountain. *Terra typica*: Chirinda Forest, Zimbabwe.

Colour variation in Marshall's Pygmy Chameleon.

GLOSSARY

Anatomical – relating to the body or body parts
Arboreal – primarily tree or shrub dwelling
Anterior – pertaining to the front portion (anteriorly – being in front of)
Antagonistic – aggressive towards other individuals
Casque – the raised bony helmet, topping the head of many chameleons
Chromatophores – specialised skin cells that contain pigments (red and yellow colours) within them
Class – a taxonomic category within the classification of life that is below **phylum** but above **order**
Cloaca – the external opening at the base of the tail, which is shared by the urinary, digestive, and reproductive systems
Conical – cone shaped in structure
Convergent evolution – adaptation to a similar environment or other conditions that produces similarity in characteristics of organisms
Cryptic (noun **crypsis**) – hidden or camouflaged
Cryptic species – species that morphologically greatly resemble each other, sometimes to the point where they cannot readily be distinguished from each other
Digits – fingers and toes
Diurnal – active during the day
Dorsal – referring to the back or the upper side of the body
Dorsal crest – the crest of modified scales that run along the spine of the back
Dorsal spines – the modified scales that form the dorsal crest
Ectotherms – animals that have metabolic rates that are too low to generate sufficient heat to maintain body temperature at high, constant levels

Endemic – native to a distinct region, found nowhere else
Epidermis – the top layer of skin
Family – a taxonomic category within the classification of life that is below **order** but above **genus**. Family names end in '*-idae*'
Fauna – animals
Flanks – the sides
Flora – plants
Genetic – pertaining to DNA or genes
Genus (pl. **Genera**) – a taxonomic category within the classification of life that is below **family** but above **species**
Granules – small non-overlapping scales (granular – referring to granule scales)
Gravid – when females contain developing eggs
Gular – pertaining to the throat or chin region
Gular crest – (see gular scales) the row of raised scales under the chin
Gular grooves – the grooves formed by the folding of the skin on the gular
Gular scales – (see gular crest) the modified scales that hang down from the throat, usually forming a row
Hemipenis (pl. **hemipenes**; adj. **hemipenal**) – in male **squamates**, the organ for copulation during reproduction
Insectivorous – feeding upon insects
Interstitial – in between
Introduction – the action of releasing a non-native plant or animal into an area where it does not naturally occur
Invertebrate – an animal that lacks a backbone, *e.g.* insects, crabs, snails
Kingdom – the highest taxonomic category within the classification of life
Lateral – pertaining to the sides or flanks
Laterally compressed – flattened sides, causing greater height than breadth

Melanophores – specialised skin cells that contain black pigment
Montane – occurring in mountains
Morphological – pertaining to the physical characters of the body (*e.g.* shape, size, scales, colour)
Nocturnal – active at night
Occipital lobe – flaps of skin that extend from the casque onto the neck
Order – a taxonomic category within the classification of life that is below **class** but above **family**
Oviparous – reproduction whereby the female lays eggs that develop into young
Phylum (pl. **Phyla**) – a taxonomic category within the classification of life that is below **kingdom** but above **class**
Pigments – compounds in the skin that give it colour
Posterior – in the back section (posteriorly – toward the back of)
Prehensile – able to grab with and to maintain a grip
Scales – thin bony plates that cover the skin of reptiles and fish
Secondary sexual characteristics – in animals, the ornate characteristics that are involved in attracting mates, such as the peacock's feathers

Sexual dimorphism – consistent and diagnosable size and/or shape differences between the sexes
Species – the taxonomic rank that is below genus; a group of related organisms that are capable of interbreeding
Species richness – the number of different species in an area
Squamates – scaled reptiles. Includes lizards and snakes.
Taxa (singular **taxon**) – a recognisable group or category (e.g. species or sub-species) in the classification of plants or animals
Taxonomy – the classification of species
Taxonomists – scientists that carry out taxonomy
Temporal crest – the bony plates on the sides of the head
Terra typica – the locality from which a species was first described
Terrestrial – living on the ground
Thermoregulation – regulation of body temperature
Tubercle – a rounded, hard protuberance. In the case of chameleons, tubercles are modified, enlarged scales.
Venation – a series of criss-crossing indentations on the skin
Viviparous – reproduction by giving birth to live young

The Cape Dwarf, fynbos form.

ONLINE RESOURCES

AdCham – http://www.adcham.com
Animal Diversity Web – http://animaldiversity.ummz.umich.edu/site/index.html
Chameleons! On-line – http://www.chameleonnews.com/index.html
EMBL Reptile Database – http://www.embl-heidelberg.de/~uetz/LivingReptiles.html
Herpetology in Southern Africa – http://www.saherps.net/

Cape Dwarf Chameleon.   Pygmy chameleon from central Mozambique.

RESOURCES

Bickel R, Losos JB (2002) Patterns of morphological variation and correlates of habitat use in Chameleons. *Biological Journal of the Linnean Society* 76, 91–103.

Branch WR (1998) *A Field Guide to Snakes and Other Reptiles of Southern Africa.* Struik, Cape Town.

Branch WR, Tolley KA, Tilbury CR (2006) A new Dwarf Chameleon (Sauria: *Bradypodion* Fitzinger 1843) from the Cape Fold Mountains, South Africa. *African Journal of Herpetology.* 55, 123-141

Cuadrado M (2001) Mate guarding and social mating system in male common chameleons (*Chamaeleo chamaeleon*) *Journal of Zoology*, London 255, 425–435.

Cuadrado M (1998) The influence of female size on the extent and intensity of mate guarding by males in *Chamaeleo chamaeleon*. *Journal of Zoology*, London 246, 351–358.

Evans SE (2003) At the feet of dinosaurs: the early history and radiation of lizards. *Biological Reviews* 78, 513–551.

Frost D, Etheridge R (1989) A phylogenetic analysis and taxonomy of iguanian lizards (Reptilia: Squamata). University of Kansas Museum of Natural History, Miscellaneous Publications 81, 65 pp.

Glaw F, Vences M (1994) *A Field Guide to the Amphibians and Reptiles of Madagascar.* 2nd Edition, 480 pp. Zoologisches Forschungsinstitut und Museum Alexander Koening, Bonn.

Glaw F, Vences M (1999) Specific distinctness and biogeography of the dwarf chameleons *Brookesia minima, B. peyrierasi* and *B. tuberculata* (Reptilia: Chamaeleonidae): evidence from hemipenial and external morphology. *Journal of Zoology*, London 247, 225–238.

Jacobsen NHG (1990) *A herpetological survey of the Transvaal.* Unpublished Ph. D. thesis, University of Natal, Durban, pp. 1–1621.

Matthee CA, Tilbury CR, Townsend T (2004) A phylogenetic review of the African leaf chameleons: genus *Rhampholeon* (Chamaeleonidae): the role of vicariance and climate change in speciation. *Proceedings of the Royal Society of London, Series B* 271, 1967–1975.

Necas P (2004) *Chameleons: Nature's*

Hidden Jewels. 2nd Edition. Chimaira Buchhandelsgesallschaft.

Necas P (2004) *Stump-tailed Chameleons: Miniature Dragons of the Rainforest.* Chimaira Buchhandelsgesallschaft.

Raw LRG (1976) A survey of the dwarf chameleons of Natal, South Africa, with descriptions of three new species. *Durban Museum Novitates* 11(7), 139–161.

Raw LRG (1978) A further new dwarf chameleon from Natal, South Africa. *Durban Museum Novitates* 11(15), 265–269.

Raw LRG (2001) Revision of some dwarf chameleons (Sauria: Chamaeleonidae: *Bradypodion*) from eastern South Africa. M.Sc. thesis, University of Natal, Pietermaritzburg.

Raxworthy CJ, Forstner MRJ, Nussbaum RA (2002) Chameleon radiation by oceanic dispersal. *Nature* 415, 784–787.

Spawls S, Howell K, Drewes R, Ashe J (2004) *A Field Guide to the Reptiles of East Africa.* Academic Press, San Diego, 543 pp.

Stuart-Fox DM, Whiting MJ (2005) Male dwarf chameleons assess risk of courting large, aggressive females. *Biology Letters* 1: 231–234.

Stuart-Fox D (2006) Testing game theory models: fighting ability and decision rules in chameleon contests. *Proc. R. Soc. B.* 273, 1555–1561.

Stuart-Fox D, Whiting MJ, Moussalli A (2006) Camouflage and colour change: antipredator responses to bird and snake predators across multiple populations in a dwarf chameleon. *Biological Journal of the Linnean Society*, 88, 437–446.

Tilbury CR, Tolley KA, Branch WR (2006) A Review of the systematics of the genus *Bradypodion* (Sauria: Chamaeleonidae) and the erection of two new genera of chameleons. ZooTaxa 1363, 23–38.

Tolley KA, Burger M (2004a) Distribution of *Bradypodion taeniabronchum* (Smith 1831) and other dwarf chameleons in the eastern Cape Floristic Region of South Africa. *African Journal of Herpetology* 53, 123–133.

Tolley KA, Burger M (2004b) Geographical distribution *Bradypodion gutturale* (Smith 1849). *African Herp News* 37, 29–32.

Tolley KA, Burger M, Turner AA, Matthee CA (2006) Biogeographic patterns and phylogeography of dwarf chameleons (*Bradypodion*) in an African biodiversity hotspot. *Molecular Ecology* 15, 781–793.

Tolley KA, Tilbury CR, Branch WR, Matthee CA (2004) Phylogenetics of the southern African dwarf chameleons, *Bradypodion* (Squamata: Chamaeleonidae). Molecular Phylogenetics and Evolution 30, 354–365.

Wager, VA (1983) *The life of the chameleon.* Wildlife Society of Southern Africa. Natal Branch, Durban.

Wolkberg Dwarf Chameleon (from Barberton).

97

INDEX

Page references in italics indicate photographs or illustrations.

A
Africa 8-9, 16-19, 28
Agama
  Southern Rock 26, *26*
Agama atra 26, *26*
Agamidae 26
Albany thicket biome 20, 24, 40
anole lizards 28
Anolis lividus 28, *28*

B
biodiversity 10, 21, 22, 43
biomes
  Southern Africa 20-25
Bradypodion 8, 9, 13, 14, 15, 16, 18, 27, 28, 38, 44, 48-85, 86
  atromontanum 75, *75*
  caffer 71-72, *71-72*
  damaranum 62-63, *62-63*
  dracomontanum 14, 54-55, *54-55*, 82
  gutturale 68-69, *68-69*
  karroicum 57
  kentanicum 60-61, *60-61*
  melanocephalum 64-67, *64-67*, 82
  nemorale 73, *73*
  occidentale 76-77, *76-77*
  pumilum 43, 50-53, *50-53*
  setaroi 74, *74*
  taeniabronchum 58-59, *58-59*
  thamnobates 70, *70*
  transvaalense 78-79, *78-79*, 82
  ventrale 56-57, *56-57*

Brookesia 8, 10, 11, 13, 32
  minima 10
  perarmata 10
  ramanantsoai 10, *10*
  stumpffi 10
  superciliaris 11
  tuberculata 10, *10*
  vadoni 11
Brookesiinae 13

C
Calumma 8, 10, 12, 13, 31
  brevicornis 12
  gastrotaenia 12
  nasuta 12, *12*
  parsonii 10, *10*, 12
Cape floristic region 20, 21
Chamaeleo 9, 13, 15, 16, 17, 28, 38, 44, 48, 49, 86-89
  arabicus 9
  calyptratus 9, 43
  caroliquarti 26
  chamaeleon 9
  dilepsis 87-88, *87-88*
  hoehnelii 9
  jacksonii 9, 43
  melleri 86
  montium *16*, 17
  namaquensis 89, *89*
  quadricornis 17
  schubotzi 9
  zeylanicus 9
Chamaeleonidae 14
Chamaeleoninae 13
Chameleon
  Arabian 9
  Asian 9

Cameroon Sailfin *16*, 17
Common Flap-necked 3, 12, 25, *44-45*, 86, 87-88, *87-88*, 89
Dwarf *see* Dwarf Chameleon
Four-horned 17
Jackson's Three-horned *16*, *36*, 43, *86*
Leaf *see* Leaf Chameleon
Mediterranean 9, 35
Meller's *86*
Mount Mulanje 19, *19*
Namaqua 3, *15*, 25, 31, 89, *89*
Nose-horned 12
Oustalet's 10, 13
Parson's 10, *10*, 12
Pygmy *see* Pygmy Chameleon
Stump-tailed 90
Usambara Two-horned *18*
Veiled *9*, 43
Werner's Three-horned 3
chameleons
  aggression 36, 37
  alien species 43
  anatomy 29-33
  avoidance 35
  behaviour 34-39
  camouflage, 31, 34
  casque 46, 49, 52
  classification 8-9
  colour change 30, 31, 46
    see also chromatophores, melanophores
  conservation 41
  cryptic coloration 10, 34
  digits 33
  distribution 8-9, 17, 18, 27, 28, 47, 86, 91
  DNA 18, 19, 52, 83
  dorsal crest 16, 47
    see also gular crest
  drinking 31, 42-43
  ears 33
  ecology 29-33
  eggs
    see also oviparity
    clutches 38, 86, 90
    hatching 38
  endangered 42, 58, 74
  evolution 26-28
  extinction 41
  eyes 33
  'false' 13
  feet 33
  fighting 35, 36, 37
  fossils 26-28
  genera 8-9
  genetics 8, 13, 18, 19, 80, 81, 85

The Spectral Pygmy Chameleon is widespread throughout equatorial Africa.

98

gestation 38, 49
granules 29, 49
growth 30
habitats 9, 12, 13, 18, 20-25, 40, 47
  arboreal 29, 31
  terrestrial 10, 29, 31, 89
  transformation 41, 42
hearing 33
horns 17, 19, 86
identification 46-47
in captivity 43
in gardens 40, 42-43, 62
incubation 38, 86, 90
interbreeding 43
life span 39, 49
litters 38
live-bearing 38
  see also viviparity
male display 37, 96
mate-guarding 35
mating behaviour 37, 38
  see also hemipenes, cloaca
maturity 38, 39, 49, 86
morphology 8, 18, 19, 46, 81, 85
  variations 12, 48
movements 34
nasal appendages 12
'New World' 28
occipital lobes 12
photographing 41, 43
predators 39, 49
prey 31
reproduction 34-39
  see also hemipenes, cloaca
scales 46
secondary sexual characteristics 49
sex determination 33
  see also hemipenal bulge
sexual dimorphism 12, 13, 49, 92
sight 33
size 12, 46, 49, 86
skin 29, 30
  see also epidermis
  shedding 30
sleep 34
snout-vent length 46
social behaviour 35, 36
species 8-9, 20, 28, 46-93
spotting 40
tail 16, 32-33, 47, 52, 53, 86
  stumpy 10, 90
taxonomy 16, 17, 19, 80, 86
teeth 31
territoriality 37
tongue 30-31
trade in 43
translocation 43

The Natal Midlands Dwarf.

'true' 13
tubercles 29, 47, 49, 52, 53
chromatophores 30
classification 14-15
  see also chameleons - classification
cloaca 33, 38, 46, 47
Comoros Islands 8, 12, 13

D

desert biome 20, 25
dorsal crest 16, 46, 47, 49
  see also gular crest
Dwarf Chameleon
  Baviaanskloof 80, *80*, 83
  Beardless 5, 80, 81, *81*, 83
  Black-headed 64-67
  Cape *3*, *6-7*, 27, *34*, 43, *47*, 50-53, *50-53*, 77
  fynbos form 53, *95*
  hybridization 52
  renosterveld form 52, 53, 77
  Drakensberg 14, *14*, 27, 54-55, *54-55*, 82
  Durban 64-67
  Eastern Cape *1*, 24, 27, *30*, 56-57, *56-57*, 68
  Elandsberg 27, 42, *42*, *49*, 58-59, *58-59*, 77
  Tsitsikamma population 58, 59
  Emerald 54, 82, *82*
  Groendal 83, *83*
  Karoo 57
  Kentani 27, 60-61, *60-61*
  Kentani Grass 60-61, *60-61*
  Knysna 27, *48*, 49, 62-63, *62-63*
  Grootvadersbosch 23, 63
  KwaZulu 27, 64-67, *64-67*
  Gilboa form 66-67
  Ixopo form 66-67
  Karkloof form 66-67
  Weza form 66-67
  Little Karoo 27, 51, 57, 63, 68-69, *68-69*, 75
  Namaqua *3*, 76-77, *76-77*
  Natal Midlands 27, 42, *46*, 65, 70, *70*, *99*
  Ngome 84, *84*
  Pondo 27, 65, 71-72, *71-72*
  Qudeni 27, 42, 73, *73*, 83
  Robertson 68-69, *68-69*
  Setaro's 27, 42, 74, *74*
  Smith's 58-59
  Southern 56-57
  Swartberg 75, *75*
  Transkei 71-72, *71-72*
  Transvaal 27, 78-79, *78-79*
  uMlalazi 85, *85*
  Western 27, *36*, 51, 68, 76-77, *76-77*
  Wolkberg *2*, 78-79, *78-79*, *97*
  Zululand *37*, *38*, 73
dwarf chameleons 27, 34, 36, 38, 44, 48-49, 50-85
  distribution 28, 48-49
  undescribed 80-85

E

endemic species 21, 22, 28, 43, 48
epidermis 30
evolution 11

F

fire 21, 25
forest biome 20, 23, 24, 40
  Afromontane 23
  Indian Ocean coastal 23, 24

99

Furcifer 8, 10, 12-13, 31
  campani 12
  minor 13
  oustaleti 10, 13
  pardalis 12
fynbos biome 20, 21, 40

G
genera 14
  see also taxa
grassland biome 25
growth
  determinant 30
  indeterminant 30
gular
  crest 46, 48, 49
  folds 46
  grooves 46, 52
  lobes 51, 52, 53
  scales 46

H
hemipenal bulge 33, 49
hemipenes 33, 38

I
Indian Ocean coastal belt biome 23, 24
insecticides 40, 42
IUCN Red Data Book 42

K
Kinyongia 8, 9, 18, 19
  adolfifriderici 18
  carpernteri 18, 19
  excubitor 9, 18
  fischeri 18, *18*, 19
  oxyrhinum 18, 19
  tavetanum 18, 19
  tenue 18, 19
  uthmoelleri 18
  xenorhinum 18, 19

L
Leaf Chameleon 10, 32
  African 90-93
  Malagasy 10, *10*, *11*
  Nosy Bé 10
  Vadon's *11*
  Warty 10, *10*
leaf-mimicry 10, 11
lizards 26, 28
Lophosaura 15

M
Madagascar 8-9, 10-13, 28
Mauritius 8
melanophores 30
Microsaura 15
miombo 25
Mount Kenya *9*, 16, 18

N
Nadzikambia 9, 18, 19
  mlanjense 19, *19*
Nama Karoo biome 20, 22
names
  common 15
  scientific 14-15

O
oviparity 38, 86, 90

P
Pygmy Chameleon *3*
  Boulenger's *90, 100*
  Chapman's *90*
  Flat-headed *29*
  Gorongoza 17, 92, *92*
  Kenya 18
  Marshall's 17, 93, *93*
  Short-tailed *18*
  Spectral 91, *98*
  Usambara Spiny 16, 17
pygmy chameleons 10, 32, 44, 90-93, 96

R
rainfall 21-25
reptiles 26
Rhampholeon 8, *9*, 10, 11, *11*, 13, 16, 17, 18, 19, 28, 32, 44, 49, 90-93
  boulengeri 17
  chapmanorum 17
  gorongosae 17, 91, 92, *92*
  marshalli 17, 91, 93, *93*
  moyeri 17
  nchisiensis 17
  platyceps 17, *29*
  spectrum *11*, 17, 91
  spinosum 16, 17
  temporalis 17
  uluguruensis 17
Rieppeleon 8, *9*, 18, 19, 32
  brachyurus 18
  brevicaudatis 18, *18*
  kerstenii 18, *18*

S
SARCA (Southern African Reptile Conservation Assessment) 42
savanna biome 20, 24, 25, 40
Seychelles 8, 13
southern Africa 16, 20, 25, 28
  biomes 20-25
species richness 20, 21
succulent Karoo biome 20, 21, 22

T
taxa 28
  see also genera
terra typica 47
thermoregulation 31
Trioceros 16-17, 38, 86
  see also Chamaeleo
  jacksonii 43
  melleri 86
  montium 17
  quadricornis 17

V
viviparity 38, 49, 86

Z
Zanzibar 8

Boulenger's Pygmy Chameleon from East Africa.